EFFECTS OF RADIATION ON SEMICONDUCTORS

DEISTVIE IZLUCHENII NA POLUPROVODNIKI

ДЕЙСТВИЕ ИЗЛУЧЕНИЙ НА ПОЛУПРОВОДНИКИ

EFFECTS
OF RADIATION
ON SEMICONDUCTORS

by
Viktor Sergeevich Vavilov

Authorized translation from the Russian
by A. Tybulewicz, B. Sc., A. Inst. P., M. I. Inf. Sc., F. I. L.

Springer Science+Business Media, LLC
1965

ISBN 978-1-4899-2722-4   ISBN 978-1-4899-2720-0 (eBook)
DOI 10.1007/978-1-4899-2720-0

The Russian text was published by the State Press for Physical and
Mathematical Literature, Fizmatgiz, in Moscow in 1963.

Вавилов Виктор Сергеевич
«Действие излучений на полупроводники»

Library of Congress Catalog Card Number 64-23245

PREFACE

The effects of electromagnetic radiation and high-energy particles on semiconductors can be divided into two main processes: (a) the excitation of electrons (the special case is internal ionization, i.e., the generation of excess charge carriers); and (b) disturbance of the periodic structure of the crystal, i.e., the formation of "structural radiation defects." Naturally, investigations of the effects of radiation on semiconductors cannot be considered in isolation. Thus, for example, the problem of "radiation defects" is part of the general problem of crystal lattice defects and the influence of such defects on the processes occurring in semiconductors. The same is true of photoelectric and similar phenomena where the action of the radiation is only the start of a complex chain of nonequilibrium electron processes. Nevertheless, particularly from the point of view of the experimental physicist, the radiation effects discussed in the present book have interesting features: several types of radiation may produce the same result (for example, ionization by photons and by charged particles) or one type of radiation may produce several effects (ionization and radiation−defect formation).

The aim of the author was to consider the most typical problems. The subjects discussed differ widely from one another in the extent to which they have been investigated. An example of a relatively intensively investigated problem is the absorption of infrared radiation by semiconductors − extensive experimental data being available at least for some substances, and theoretical interpretations being available for the majority of cases. An example of an important but neglected problem is the formation and physical nature of the radiation defects in semiconductors. The results of the studies of radiation effects in semiconductors are not only of scientific value but are also essential to the successful solution of several important practical problems, such as:

a. the direct conversion of solar and nuclear radiation energy into electrical power;

b. the recording of weak infrared radiation fluxes;

c. the design of new electromagnetic radiation sources (masers and lasers) using semiconductors;

d. the counting and determination of the energy and total flux of fast particles and gamma-ray quanta;

e. the application of semiconductor electronics to nuclear power.

The solution of each of these problems necessarily involves subjects far removed from the physics of semiconductors. In recent years, several monographs and reviews of Soviet and foreign authors, presenting the advances made toward the solution of these problems, have appeared in the USSR [1-9].

The contents of the present book reflect to a considerable extent the interests of a team of workers studying the effects of radiation on semiconductors at the P. N. Lebedev Physics Institute of the USSR Academy of Sciences and the author's experience of lecturing to the senior students of the physics faculty of the M. V. Lomonosov Moscow State University.

The author is deeply grateful to B. M. Vul, Corresponding Member of the USSR Academy of Sciences, for his interest in the work on the effects of radiation on semiconductors and for his numerous valuable comments. The author is also very grateful to V. S. Vinogradov, A. A. Gippius, V. D. Egorov, and S. M. Ryvkin for their criticism of and comments on the book when in manuscript.

CONTENTS

Chapter I

ABSORPTION OF LIGHT BY SEMICONDUCTORS

Chapter II

PHOTOIONIZATION AND PHOTOCONDUCTIVITY IN SEMICONDUCTORS

Chapter III

IONIZATION OF SEMICONDUCTORS BY
CHARGED HIGH-ENERGY PARTICLES

Chapter IV

RADIATIVE RECOMBINATION IN SEMICONDUCTORS;
POSSIBILITY OF THE AMPLIFICATION AND GENERATION
OF LIGHT USING SEMICONDUCTORS

Chapter V

CHANGES IN THE PROPERTIES OF SEMICONDUCTORS
DUE TO BOMBARDMENT WITH FAST ELECTRONS,
GAMMA RAYS, NEUTRONS,
AND HEAVY CHARGED PARTICLES

Part A. Theoretical Interpretations
of the Process of Radiation-Defect Formation

Chapter I

ABSORPTION OF LIGHT BY SEMICONDUCTORS

§1. Optical Constants of Semiconductors and Methods of Determining Them

The absorption of light* by semiconductors and dielectric crystals may be accompanied by photoionization, i.e., the generation of excess densities of electrons and holes in the conduction and valence bands, and by electron transitions to excited states. It may also be accompanied by other processes (the excitation of lattice vibrations, interband electron transitions, etc.), but from the point of view of the problems to be discussed later, it is the former processes, especially photoionization, that are particularly interesting.

Electrons in a crystal may be divided into the following groups depending on the nature of their interaction with electromagnetic radiation:

  a. electrons in the valence band;

  b. charge carriers (electrons in the conduction band and holes in the valence band);

  c. electrons localized at defect or impurity levels;

  d. electrons of the inner shells of atoms.

The optical properties of a material are represented by the refractive index n and the absorption index $\varkappa$, which is also known as the extinction coefficient. We shall restrict ourselves to a discussion of nonmagnetic isotropic media whose permittivity ε and conductivity σ are scalars. The value of ε is found from the expression

$$\varepsilon = n^2 - \varkappa^2 = 1 + 4\pi\chi. \qquad (1.1)$$

The susceptibility χ is related to polarization. Usually,

---

*Here, the term "light" represents electromagnetic radiation over a wide range of wavelengths.

2 ABSORPTION OF LIGHT BY SEMICONDUCTORS [Ch. I

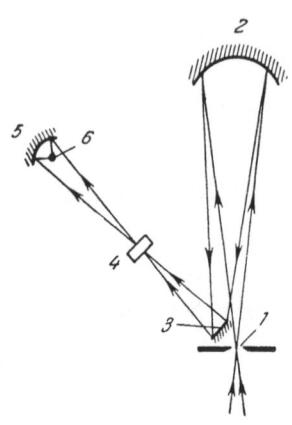

Fig. 1. Measurement of optical transmission in a sample with plane surfaces. 1) Monochromator slit; 2) spherical mirror; 3) plane mirror; 4) sample; 5) elliptical mirror; 6) radiation receiver.

$$\varepsilon \approx n^2, \quad \text{since} \quad n^2 \gg \varkappa^2. \qquad (1.2)$$

The conductivity σ, which governs the absorption of energy, is given by the formula

$$\sigma = n\varkappa\nu, \qquad (1.3)$$

where ν is the frequency.*

The experimentally determined absorption coefficient α is related to the absorption index $\varkappa$ by the expression

$$\varkappa = 4\pi\varkappa\frac{\nu}{c} = 4\pi\varkappa\bar{\nu}, \qquad (1.4)$$

where $\bar{\nu} = \nu/c$ is the wave number in cm^{-1}.

Data on the optical constants n, α, and κ are obtained by investigating the transmission of light by the test material, or its reflectivity R. A simple method for making optical measurements on samples of crystals with polished surfaces is shown in Fig. 1. The measured quantity is the transmission T = I/I_0, i.e., the ratio of the intensities of the incident I_0 and transmitted I light beams. For monochromatic radiation of wavelength λ

$$T = \frac{I}{I_0} = \frac{(1-R)^2 + 4R\sin^2\psi}{e^{ad} + R^2 e^{-ad} - 2R\cos 2(\varphi+\psi)}, \qquad (1.5)$$

where d is the sample's thickness, and the values of the angles φ and ψ are given by the formulas:

$$\varphi = \frac{4\pi nd}{\lambda}, \quad \psi = \tan^{-1}\frac{2\varkappa}{n^2+\varkappa^2+1}.$$

The term $2R\cos 2(\varphi+\psi)$ represents the interference in a plate-shaped sample. The formula (1.5) also allows for multiple reflection from the surfaces. If we use samples of sufficient thickness or light covering a wide range of the spectrum $\Delta\lambda$, we can avoid interference and use the simple formula

* The conductivity σ depends on the optical frequency ν and, in general, it is not equal to the conductivity σ_0 at zero or low frequencies.

$$T = \frac{I}{I_0} = \frac{(1-R)^2}{e^{\alpha d} - R^2 e^{-\alpha d}}. \tag{1.6}$$

The above formula allows for multiple internal reflections, which are important when the transparency and reflectivity are high. The reflectivity of a clean surface, R, for normal incidence is given by the formula

$$R = \frac{(n-1)^2 + \varkappa^2}{(n+1)^2 + \varkappa^2}. \tag{1.7}$$

Under real conditions, the reflectivity may depend strongly on the state of the surface, in particular on the presence of thin oxide films. To determine the value of α, it is convenient to eliminate R by carrying out measurements on samples of different thickness but having the surface treated in the same way. The value of n may be obtained from measurements of the reflectivity or of the interference in thin plates [2].

When the absorption of light by crystals is sufficiently weak, the value of the refractive index, which really represents the volume properties, is found by shaping a given material into a prism and measuring the deviation of a light beam which passes through it [3].

Table 1 gives values of n and ε for elements of group IV of Mendeleev's table and some intermetallic compounds of the A_3B_5 type.

It follows from Eq. (1.7) that in the absence of absorption the transmission of a plane-parallel plate is governed by the value of n.

§ 2.   "Intrinsic" Optical Absorption Band (Fundamental Band)

The existence in all semiconductors of a wide spectral region of very intense absorption, limited on the long-wavelength side by a sharp edge, is due to the fact that the absorption of photons of sufficiently high energy is accompanied by electron transitions from the valence to the conduction band.

In the case of covalent crystals or crystals with weak ionic binding, light of frequency $\nu < (E_g/h)$, where E_g is the "thermal" width of the forbidden band (gap), passes through pure crystals without causing photoionization.

TABLE 1.   Refractive Indices n and Permittivities
ε of Some Semiconductors

Substance	n	ε	Substance	n	ε
Diamond	2.417	5.9	InSb	3.988•	15.9
Si	3.446 •	11.8	GaP	2.97†	8.4
Ge	4.006•	16.0	GaAs	3.348•	11.1
InP	3.37†	10.9	GaSb	3.748•	14.0
InAs	3.428•	11.7	AlSb	3.188•	10.1

•Prism method.
†Reflection data [4].

The readily observed increase in the absorption coefficient
for photons of energies $h\nu > E_g$ allows us to estimate the value
of the forbidden bandwidth. The nature of the absorption increase
with increasing photon energy, i.e., the shape of the absorption
band edge, is governed by the electron energy-band structure of
the semiconductor. Absorption processes competing with the "intrinsic" absorption and the difficulty of determining exactly small
values of α (beginning with fractions of 1 cm^{-1} or less) usually
prevent us from obtaining very exact values of E_g from the data
on the absorption of light by crystals. On the other hand, an approximate value of E_g obtained by this method is reliable and the
method itself is valuable, because of its simplicity, in the initial
studies of new semiconducting materials (an accurate optical method for determining E_g from the fine structure of the recombination radiation spectra will be described later, in Chapt. IV).

It is known that in ionic crystals the thermal and optical forbidden bandwidths are different. The optical excitation energy
in these crystals is found to be greater than the thermal excitation energy [5]. This condition can be explained qualitatively
using the Franck-Condon principle, according to which the excess energy of a system which has absorbed a photon is transformed into the energy of lattice vibrations, in a time considerably
longer than the duration of the act of absorption.

The expression for the absorption coefficient, corresponding
to an electron transition from a state i in the valence band to a
state f in the conduction band without phonon participation, has
the form

$$\alpha = \frac{c}{\nu} |P_{if}|^2 N(h\nu),\qquad\qquad(1.8)$$

where c is a constant representing the medium, and N(hν) is the density distribution of the final states over an interval of unit energy. The matrix element representing transitions of this type is

$$P_{if} = -i\hbar \int \psi_f^* e_k \,\text{grad}\, \psi_i \, d\tau, \tag{1.9}$$

where e_k is the polarization vector of radiation with the wave vector **k**; $i = \sqrt{-1}$. Following Bloch, we can write the wave functions of electrons in the following form:

$$\psi_{kn} = e^{ikr} U_{kn}(r), \tag{1.10}$$

where U_{kn} are periodic functions with the same period as the lattice, we find that P_{if} vanishes at any point where the following selection rule is not satisfied

$$k_i + k = k_f. \tag{1.11}$$

Since the wave vectors of an electron in its initial and final states are much greater than the wave vector of a photon, the above selection rule can be expressed also as

$$k_i \approx k_f. \tag{1.12}$$

Thus, in agreement with the law of conservation of momentum, only the "vertical" transitions without any change in the wave vector are allowed.

A careful study of the fundamental band edge of germanium single crystals, carried out on samples whose thickness was in some cases a fraction of a micron, allowed us to detect the structure shown in Fig. 2 [6]. By the time these experimental results were obtained, it had been shown – by the cyclotron resonance method – that the bottom of the conduction band in germanium crystals did not correspond to the electron wave vector **k** = 0. The band structure of germanium was calculated theoretically by Herman, whose results are shown schematically in Fig. 3 [7]. In accordance with the above selection rule and Herman's data, the energy threshold for the vertical transitions should correspond to the frequency

$$\nu' = \frac{1}{h} (E_c' - E_v),$$

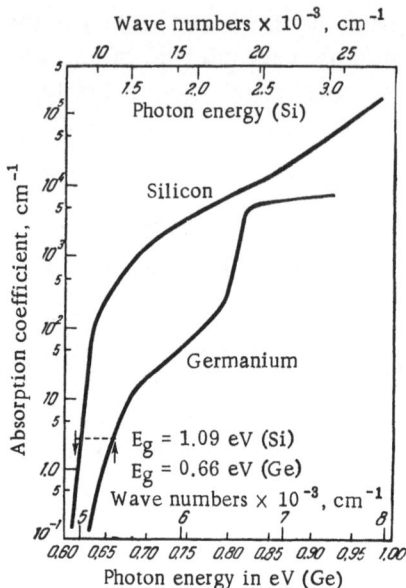

Fig. 2. Edge of the fundamental optical absorption band of germanium and silicon single crystals at 300°K [1].

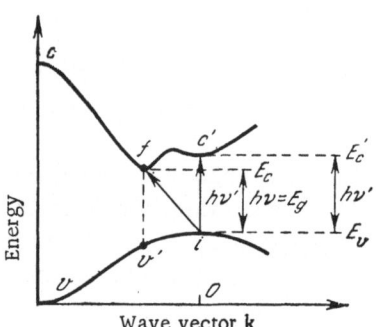

Fig. 3. Energy-band structure of a germanium crystal: v valence band; c conduction band. The complex structure of the valence band is not shown.

which is considerably higher than

$$\nu = \frac{E_g}{h} = \frac{1}{h}(E_c - E_v).$$

However, it is evident from the experimental curve (Fig. 2) that the absorption of a pure germanium single crystal increases strongly at $\nu \approx E_g/h$. To explain this fact, J. Bardeen, F. J. Blatt, and L. H. Hall [8] suggest that in the region of photon energies insufficient for vertical transitions, electron transitions to the conduction band still occur because the selection rule of Eq. (1.12), which should be strictly obeyed in an ideal periodic crystal, is relaxed due to the interaction of electrons with phonons.

Returning to the energy-band structure of Fig. 3, we must follow Herman [7] in assuming that an electron is excited optically from a state i to c' and is then transferred from c' to f emitting or absorbing a phonon. As a result of this, the electron wave vector changes considerably and the whole process can be considered as a "nonvertical" transition from i to f with the absorption of a photon $h\nu \approx E_g$.

Since each of the processes shown in Fig. 3 may involve the emission or absorption of a phonon, the matrix element which gives the transition probability becomes

$$P_{if} = \sum_{m=1}^{4} \frac{P_{im}P_{mc}}{E_m - E_{v'}}, \tag{1.13}$$

where m denotes the intermediate state (c' or v'). The quantities $P_{ic'}$ and $P_{v'f}$ are found from Eq. (1.9). For $P_{iv'}$ and $P_{c'f}$, we have the selection rule

$$k_i = k_f + q,$$

where q is the wave vector of a phonon. Assuming that the phonon matrix elements are constant, and that $E_{v'} - E_i \approx E_{c'} - E_f$, Bardeen, Blatt, and Hall [8] have shown that the wavelength dependence of the absorption coefficient for nonvertical transitions should be

$$\alpha_i \sim (h\nu - E_g \pm h\nu_q)^2 \tag{1.14}$$

in the case when transitions are possible for $k = 0$ ($h\nu_q$ represents the phonon energy). In the region of photon energies sufficient for direct (vertical) transitions, beginning with the threshold energy E_g', the absorption coefficient α_d increases, according to the theory, as follows

$$\alpha_d \sim (h\nu - E_g')^{\frac{1}{2}}. \tag{1.15}$$

The considerable difference between the last two expressions can be sometimes used to analyze the experimental data on the form of the absorption band edge and to explain the nature of electron transitions.

Up to now, the absolute values of the photon absorption cross sections * within the fundamental band have not been calculated exactly. We may justifiably assume that when the absorption coefficient is greater than 10^4 cm^{-1}, allowed interband transitions take place in a semiconductor crystal [1].

Figure 4 gives the dependence of the absorption coefficients of germanium and silicon on the difference $\zeta - \zeta_t$ (ζ is a wave number in general, and ζ_t are the wave numbers corresponding to the forbidden bandwidth, i.e., $\zeta_t = E_g/hc$). From the curves of Fig. 4, it is evident that the power exponent n in the dependence $\alpha \sim (\zeta - \zeta_t)^n$ at not too high values of α is close to 2.5 for

* We mean here the ratio α/N, where N is the number of atoms per cm^3 of a crystal.

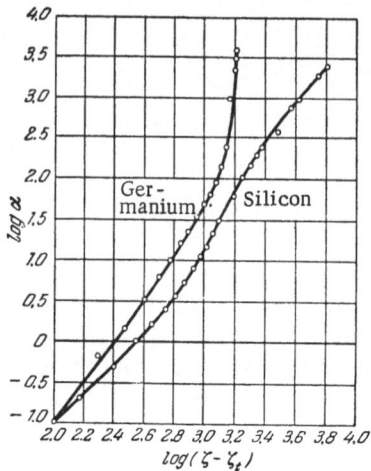

Fig. 4. Dependence of the logarithm of the absorption coefficient on the logarithm of the difference $(\zeta - \zeta_t)$ for germanium and silicon.

germanium and 2.0 for silicon. Both these values are several times greater than the power exponent n = 0.5 which one would expect for vertical transitions. The conduction band of silicon, like that of germanium, has a maximum corresponding to $k \neq 0$ (Fig. 2). Thus the experimental data for germanium and silicon are in agreement with the theory of nonvertical transitions.

The steep rise in the absorption coefficient of germanium crystals near 0.81 eV (Fig. 2) is explained by the fact that this energy corresponds to the threshold for vertical transitions. Comparison of this value with the data on the band structure of germanium, obtained by the other methods mentioned above, allow us to conclude that the minimum in the conduction band corresponding to the origin of coordinates (0, 0, 0) lies above the minimum on the [1, 1, 1] axis. This conclusion regarding the difference of the minimum energies necessary for nonvertical and vertical transitions in germanium is also confirmed by the data on the spectrum of the direct recombination radiation of Ge obtained by J. R. Haynes, which will be discussed in Chapt. IV.

The earlier data on the absorption band edge of Si, obtained by the method of reflection coefficients [9], suggested that the threshold energy for vertical transitions was close to 1.8 eV. However, a direct determination of α, as well as a careful study of the recombination radiation spectra, did not confirm this conclusion. According to the measurements of W. C. Dash and R. Newman [6], the vertical separation between the conduction and valence bands at k = 0 amounts to 2.5 eV. An analysis of the form of the absorption band edge of PbS, PbTe, and PbSe single crystals, carried out recently by Scanlon [10], who also used the above formulas, gave reliable data on the forbidden bandwidths of these very interesting and important (in infrared technology)

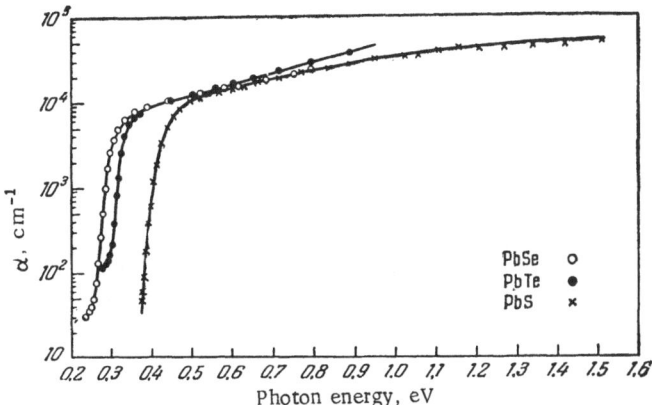

Fig. 5. Edge of the fundamental absorption band of the semiconducting compounds PbS, PbSe, and PbTe [10].

semiconductors. As in the case of germanium and silicon, in
earlier work on the absorption of PbS, two different methods were
used for the absorption coefficients ranging from 1 to 10^6 cm^{-1}:
the transmission was measured at low values of α and the reflectivity at high values.

The results of these measurements were found to differ. To
obtain fully reliable data, the following method was adopted. Using
the tendency of PbS-type crystals to cleave along the (100) planes,
a series of single-crystal samples was prepared from several
microns to 1 mm thick, and measuring more than $600 \times 50 \mu$
across. Then, having improved the focusing part of an infrared
spectrograph, it was possible to measure the transmission throughout the absorption-edge region of PbS, PbSe, and PbTe (Fig. 5).

The plots shown in Fig. 6 were used to analyze the form of
the absorption edge. These plots show clear linear portions corresponding to values $\alpha > 3000$ cm^{-1}, the extrapolation of which
gives the value of E_g for vertical (direct) transitions (cf. Table 2).
The results of Scanlon should be regarded as more reliable than
the earlier work reporting much higher values of E_g for PbS and
similar compounds [11].

If we plot the dependence of $\alpha^{1/2}$ on $E = h\nu$ (Fig. 7), we find
that in the $\alpha < 3000$ cm^{-1} region there is an approximately linear
part which may be related to nonvertical electron transitions. The
values of $h\nu = E_g$ for nonvertical transitions   extrapolated to $\alpha = 0$

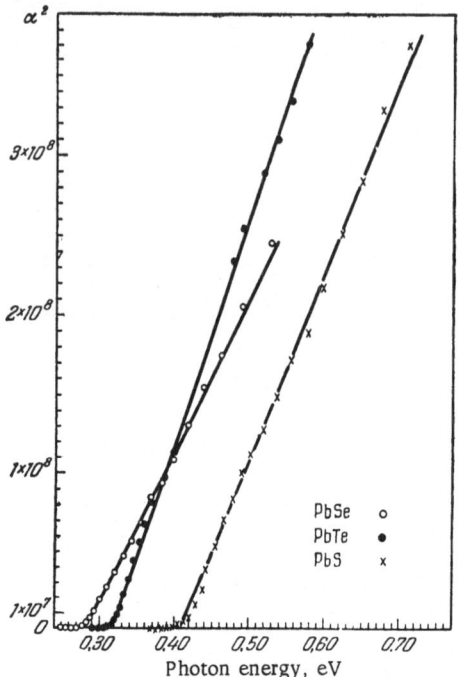

Fig. 6. Part of the fundamental absorption band edge
of the compounds PbS, PbSe, and PbTe, corresponding
to direct transitions [10].

are also listed in Table 2. It is worth noting that in compounds
of the PbS type the values of E_g for direct and indirect transitions
are, in contrast to Ge, close to one another. According to published work, the Debye temperatures of these compounds lie in the
region 150-200°K [12]. Therefore, the small differences in the
values of E_g observed experimentally may be explained by the
participation of lattice vibrations without the additional assumption that the minimum in the conduction band does not lie above
the maximum in the valence band.

For quite a long time, there was no agreement as to how to
determine the "optical" width of the forbidden band E_{g0}. * Ac-

* See, for example, the monograph of T. S. Moss "Optical properties of semiconductors" [17].

TABLE 2. "Optical" Width of the Forbidden Band
from Analysis of the Absorption-Band Edge

Substance	E_g at 300°K in eV	
	Vertical transitions	Nonvertical transitions
PbS	0.41	0.37
PbSe	0.29	0.26
PbTe	0.32	0.29

cording to present ideas, in those cases where the form of the absorption band is known, for example, Ge or PbS, the value of E_{g0} is determined by extrapolating the theoretical dependence $\alpha = f(h\nu)$ to zero values of α. The value of E_{g0} can be found even more precisely in those cases when the fine structure of the absorption or emission spectra has regions related to the emission and absorption of phonons [18].

However, for those semiconductors for which a direct comparison with theory cannot yet be made and the value of E_{g0} is found approximately from the position of that part of the absorption edge where α varies most strongly with wavelength, the error in the value of E_{g0} does not usually exceed 0.03 eV for single-crystal samples. Data on the optical width of the absorption band are given for several semiconductors in Table 3.

Optical measurements in the region of the absorption band edge may give important results on degenerate semiconductors. An interesting example is indium antimonide, InSb, the constant energy surfaces of which are nearly spherical. In InSb crystals, the absorption band edge corresponds to vertical electron transitions. The forbidden bandwidth of intrinsic InSb at 300°K amounts to 0.175 eV [13]. It was found that in n-type InSb with high electron density the fundamental absorption band edge was strongly displaced toward short wavelengths. Allowing for the small effective electron mass ($0.015 m_e$) E. Burstein suggested that this effect was due to the filling of the energy levels near the bottom of the conduction band. Since the effective electron mass is small, the lower part of the conduction band is filled even at densities of the order of 10^{18} cm^{-3}, and electrons from the valence band may move only to the states much higher than the bottom of the conduction band. According to Burstein, the position of the

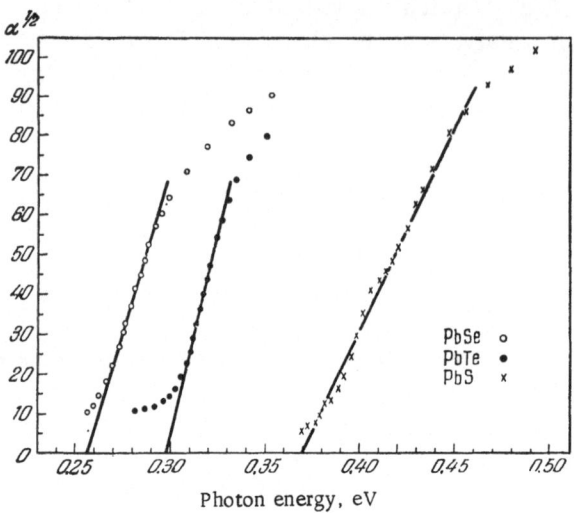

Fig. 7. Part of the fundamental absorption band edge of the compounds PbS, PbSe, and PbTe, corresponding to indirect transitions [10].

fundamental band edge $E = E_g + \Delta E$ is a function of the electron density in the conduction band n

$$\Delta E = \left(1 + \frac{m_n^*}{m_p^*}\right)(E_F - \gamma kT), \qquad (1.16)$$

where m_n^* and m_p^* are the effective masses of electrons and holes, respectively; E_F is the Fermi energy and γ is the parameter which depends on the population of the states near the minimum of the conduction band. According to calculations of H. Y. Fan and W. Kaiser [1]

$$\gamma = \ln \frac{\alpha_0 - \alpha}{\alpha}, \qquad (1.17)$$

where α_0 is the absorption coefficient corresponding to an unfilled conduction band. The curve of Fig. 8 was plotted using the last two formulas. To obtain the best agreement with experiment, it was necessary to use an effective mass $m_n^* = 0.03\, m_e$, and not $0.015\, m_e$. In this connection, it was suggested [4] that if the Fermi level lay above the edge of the conduction band the effective electron mass increased.

TABLE 3.   Width of the Forbidden Band of Some Semiconductors
Determined by the Optical Method [4]

Substance	E_{g0} at 300°K, eV	E_{g0} at 0°K, eV	Substance	E_{g0} at 300°K, e V	E_{g0} at 0°K, eV
Si	1.09	1.14	GaP	2.24	2.4
Ge	0.66	0.75	GaAs	1.45	1.53
InP	1.25	1.34	GaSb	0.70	0.80
InAs	0.35	0.45	AlSb	1.60	1.70
InSb	0.175	0.25			

§ 3.   Effect of Temperature, Pressure, Electric and Magnetic Fields on Optical Absorption in the Fundamental Band

A.   Effect of Temperature and Pressure

Experimental data, from both electrical and optical measurements, indicate that the forbidden bandwidth of semiconductors
depends on temperature. This dependence, which within certain
limits is nearly linear, is related mainly to the temperature dependence of the unit cell dimensions [14]. Using the theory of the
formation of allowed energy bands from atomic levels when atoms
approach one another, it is possible to explain qualitatively the
narrowing of the forbidden band with increase in temperature
(which is characteristic of germanium, silicon, and many other
semiconductors) as well as the converse effect (which occurs,
for example, in PbS-type crystals). A theoretical analysis of
the dependence of the forbidden bandwidth on temperature and
pressure is based on the fact that a deformation of the lattice
changes its potential by ΔU [19]. The crystal potential changes
also due to thermal vibrations. When the forbidden bandwidth
changes due to the compression of a crystal by external pressure,
the effect is solely due to the change in the lattice constants. This
is also true of that part of the absorption band edge which corresponds to those vertical electron transitions which appear without a change in the wave vector **k**.

Temperature variations affect not only the dimensions of a
unit cell but also the spectrum of thermal vibrations, and this
may produce additional changes.

This phenomenon was considered by H. Y. Fan [1, 15], who
obtained a satisfactory agreement with the experimental results

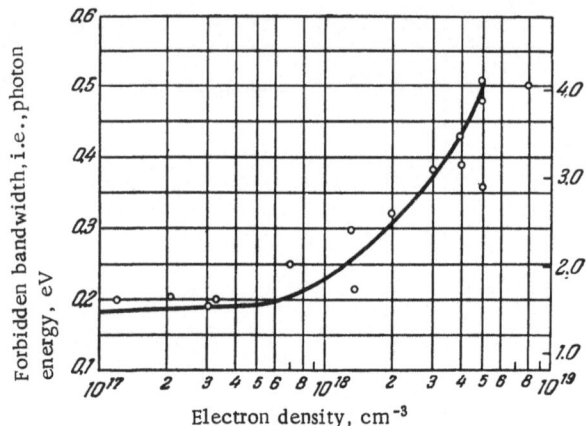

Fig. 8. Edge of the fundamental absorption band of InSb
as a function of the electron density in the conduction
band. The continuous curve represents theoretical calculations.

for germanium and silicon crystals. The most detailed data on
the temperature dependence of the absorption band edge on temperature have been obtained for germanium [16]. The data of H. Y.
Fan and other workers are presented in Fig. 9; those of W. C.
Dash and R. Newman [6] in Fig. 10. From these two sets of results, it is evident that, in general, the curves corresponding to
different temperatures may be obtained from one another by appropriate displacements along the horizontal and vertical axes.
The absorption at long wavelengths, which increases steeply at
high temperatures, is explained by the interaction of radiation
with charge carriers, which is not related to interband electron
transitions.

More accurate data on the dependence of E_{g0} on T at low temperatures, where the carrier absorption (superimposed on the
fundamental band) is quite weak, are given for germanium [20]
in Fig. 11. In the region of the linear dependence of E_g on T
(above 150°K), the value of $\partial E_{g0}/\partial T$ for germanium agrees with
that determined from the change in the carrier density. It is
evident from the curve in Fig. 11 that the rate of variation of E_{g0}
with temperature decreases at low temperatures.

The change in the forbidden bandwidth due to the compression
of a crystal was investigated for germanium and silicon at con-

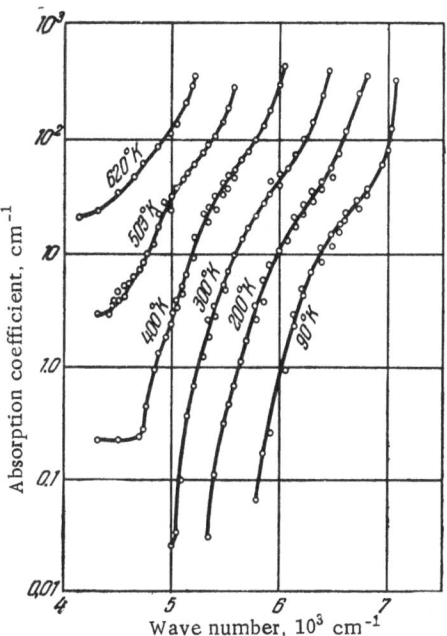

Fig. 9. Variation in the position and shape of the
fundamental adsorption band edge of germanium
with temperature [6].

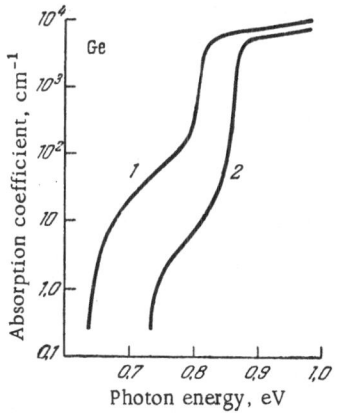

Fig. 10. Edge of the absorption band
of germanium at 300°K (1) and 77°K
(2).

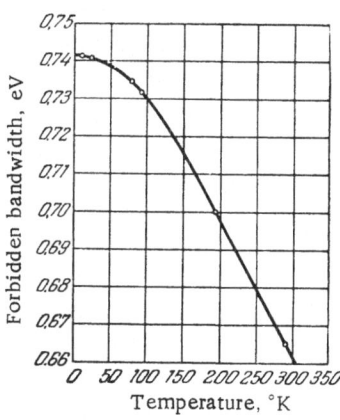

Fig. 11. Temperature dependence of
the "optical" forbidden bandwidth of
germanium [20].

stant temperature [21]. The experimental values of $(\partial E_g/\partial P)_T$, determined from the displacement of the fundamental absorption edge, were close to 5×10^{-12} eV $\cdot$ cm^2 $\cdot$ g^{-1}. However, the sign of this quantity was positive for germanium and negative for silicon.

B. Effect of an Electric Field on the Edge of the Fundamental Optical Absorption Band

The region of strong "intrinsic" absorption of crystals has a sharp threshold on the long-wavelength side, which corresponds to the minimum distance between the energy bands. Quanta whose frequencies are lower than $\nu_0 = E_{g0}/h$ cannot be absorbed since their energy is insufficient to transfer an electron from the valence to the conduction band. The application of an external electric field alters the absorption probability by spreading the edge of the fundamental absorption band. Therefore, an electron with momentum **p** in an external field cannot be given any definite energy since the probability of finding it is the same throughout the region $L \approx E_g/qE$ of a crystal but its potential energy may have different values. Conversely, if the potential energy is strictly fixed, i.e., if the electron is localized in a certain cell of a crystal, its momentum may take any value within the limits of a given band and its energy may again be indeterminate.

In order to find as accurately as possible the bottom of a band near a given point in space x it is necessary to form a packet, Δx wide, of the wave functions of the states close to the edge of this band, retaining only the states whose momentum does not differ by more than Δp from the momentum corresponding to the bottom of the band. The parameters Δp and Δx should be defined so that the energy indeterminacy

$$\Delta E \sim \frac{(\Delta p)^2}{m^*} + q\mathscr{E}\,\Delta x \qquad (1.18)$$

is minimal (m* is the effective electron mass, q is the electronic charge, and $\mathscr{E}$ is the field intensity). From the condition for the minimum of this expression, using the indeterminacy relationship $\Delta p \cdot \Delta x \geq \hbar$, it follows that

$$\Delta E_{min} \approx \sqrt[3]{(q\mathscr{E})^2 \frac{\hbar^2}{m^*}}. \qquad (1.19)$$

Consequently, in an electric field, the edge of an energy band – and therefore the edge of an absorption band – should spread to a value of the order of ΔE_{min}. In particular, the absorption of quanta whose energy is less than E_{g0} may become possible. Within certain limits, this spreading may be interpreted as the shift of the absorption "threshold" toward longer wavelengths.

A rigorous examination, undertaken by L. V. Keldysh [22], of the problem of the absorption of light by a crystal placed in a uniform electric field confirms these qualitative conclusions * (see also [24]). In fields of the order of 10^5 V/cm, the quantity $\Delta E_{min} = h \Delta \nu_g$ may reach 0.01 eV, which is considerably greater than the value of the possible Stark effect which, as shown by F. F. Vol'kenshtein, also produces a shift of the fundamental absorption edge [25].

V. S. Vavilov and K. I. Britsyn [26] carried out experiments to detect the influence of a strong electric field on the absorption of light in silicon single crystals. It was not possible to employ the usual single crystals, prepared by pulling from the melt or by zone-melting, because their resistivities were too low. They used silicon of about 10^{11} Ω · cm resistivity at the test temperature (T = 100°K); this resistivity was obtained by irradiation in a reactor with an integral fast-neutron dose of the order of 10^{18} neutrons/cm^2. From earlier investigations of the optical properties of such irradiated silicon, it was known that near the absorption edge, a new band, due to defects, appeared close to 1.8 μ, and the transmission fell somewhat in the immediate vicinity of the fundamental absorption edge [27, 28]. However, this change in the optical properties was due to local disturbances and not to a change in the band structure itself, so that, using sufficiently thin samples, it was possible to carry out measurements at wavelengths where the absorption was mainly due to interband transitions.

The experimental layout is shown in Fig. 12a. A sample in the form of a rectangular slab was placed between metal electrodes, one of which was at the temperature of liquid nitrogen. Modulated monochromatic light passed through the whole crystal, the ends of which were polished, and fell on a cooled PbS photoresistor. The alternating voltage which developed across the

---

* Similar results were obtained independently by W. Franz [23].

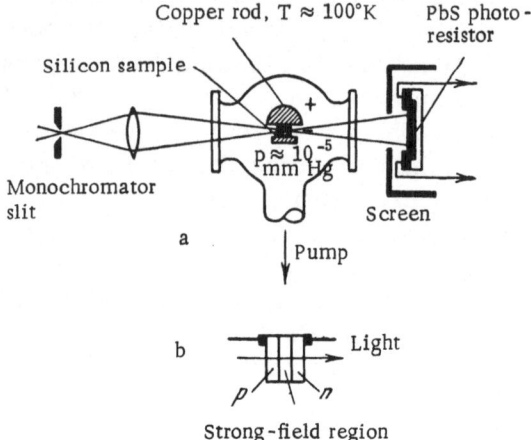

Fig. 12. Layout for measurements of the electro-
optical effect in a uniform semiconductor crystal (a)
and in a crystal with a p-n junction (b).

photoresistor was fed to a narrow-band amplifier with its output
connected to a pointer instrument.

In the spectral region corresponding to transitions between
the valence and conduction bands with phonon participation, a considerable displacement of the absorption-band edge was detected
(Fig. 13). Figure 13b shows that in the wavelength region 0.8-
0.9 μ the application of an electric field strongly increased the
absorption coefficient. The value of the observed shift $\Delta\lambda$ and
the dependence of this shift on the field intensity were in agreement with the theoretical predictions. The magnitude of the effect and the small change in the conductivity due to the excitation
of nonequilibrium carriers suggest that the observed phenomenon
is the intrinsic effect of the field and is not due to the absorption
by carriers. The measurements were carried out under steady-
state conditions ($\mathscr{E}$ = const) but the nature of the effect forced
us to assume that it was a rapid change and, therefore, using a
simple device, similar to the one shown in Fig. 12, it was possible
to modulate the light of 0.8-1.0 μ wavelength with very high frequencies. Recently, K. I. Britsyn [29] measured the optical
transmission of Si crystals to which a high frequency field was
applied. It was found that the delay in the signal, corresponding

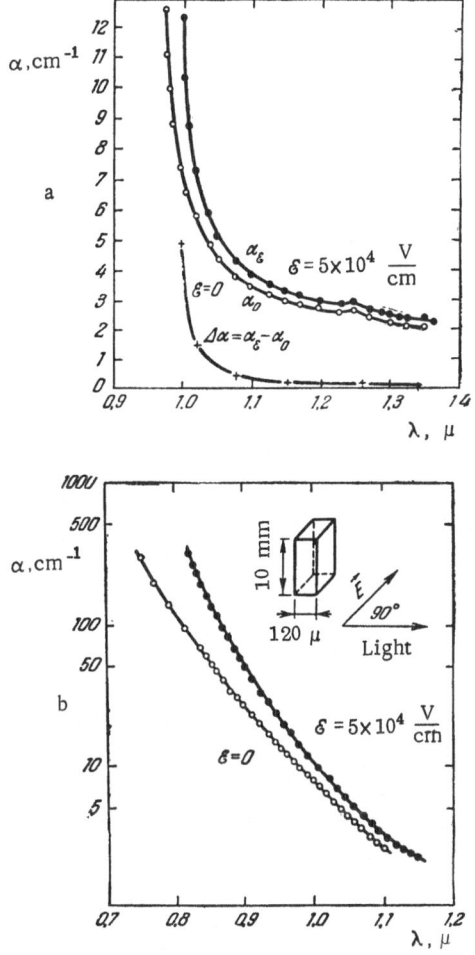

Fig. 13. Electro-optical effect in silicon [26]. Spectral resolution $\Delta \lambda \approx 0.005 \ \mu$.

to the additional absorption, with respect to the field $\mathscr{E}$ did not exceed 2×10^{-8} sec.

C. Effect of a Magnetic Field on the Fundamental Absorption Band

Measurements of the absorption in crystals of germanium, InSb, and InAs, placed in a strong magnetic field, showed that a well-defined system of periodically distributed absorption maxima ap-

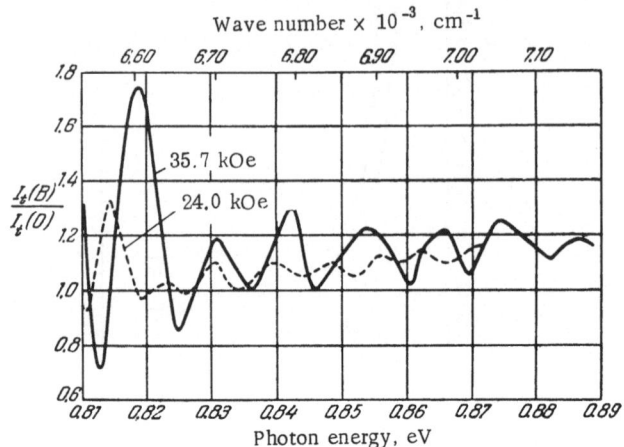

Fig. 14. Influence of a magnetic field on the optical absorption in the fundamental band (Landau levels). The quantities $I_t(B)$ and $I_t(0)$ are, respectively, the relative intensities of the transmitted light with and without magnetic field B.

peared within the fundamental band, near its edge (Fig. 14). The positions of these maxima varied with field intensity. This effect was called the "oscillatory magnetoabsorption" by the workers who detected it in germanium [30].

The absorption maxima appearing on the application of a magnetic field correspond to electron transitions between the Landau levels in the valence and conduction bands. These discrete levels appear due to the accumulation of the levels in the continuum around certain positions in the bands, this accumulation being enhanced by the application of the field. For simple parabolic energy bands, the Landau level positions are given by the expression

$$E_L = \left(n + \frac{1}{2}\right) h\nu_c + \frac{h}{2\pi} \frac{k_z^2}{2m^*}, \tag{1.20}$$

$$\nu_c = \frac{q_B}{2\pi m^*}, \tag{1.21}$$

where n is the quantum number (the level number); ν_c is the cyclotron frequency; k_z the projection of the wave vector along a direction parallel to the magnetic induction vector B [31]. From

the above expression, it follows that the change of the forbidden
bandwidth in a magnetic field is

$$\Delta E_{g0} = \frac{1}{2} h (\nu_{cn} + \nu_{cp}),$$ (1.22)

where ν_{cn} and ν_{cp} are the cyclotron frequencies of electrons and
holes, respectively.

Investigations of the "oscillatory magnetoabsorption" can be
used to determine the reduced effective mass of carriers. In
principle, it should also be possible to determine the minimum
photon energy necessary to produce a direct (vertical) transition
between the bands. For this purpose, it is necessary to extra-
polate to zero magnetic field the positions of the magnetoabsorption effect maxima. Such an extrapolation was carried out for
germanium by S. Swerdling et al. [30], and it gave a value identical with E_{g0} for vertical transitions found by other methods (see,
for example, [6]).

However, later, more accurate measurements carried out
on germanium crystals showed additional absorption maxima related to complex band structure and exciton formation, which made
it difficult to interpret experiments of this type [4].

§ 4.   Absorption of Light Accompanied
by Exciton Formation

The concept of an excited state of the electron system of a
crystal, not connected with localized centers, was put forward
by Ya. I. Frenkel', who suggested the term "exciton" to denote
such a state. One of the experimental facts explained by the exciton hypothesis is the existence of bands next to the fundamental
absorption edge. The absorption of light in these additional bands
is not accompanied by photoconductivity. An exciton may be regarded as an excited state propagated from one crystal cell to another, or as a system consisting of an electron and a positive hole,
similar to the hydrogen atom. It follows, first, that the motion
of an exciton in a crystal does not produce electric current and,
secondly, that to form an exciton less energy is required than to
generate a "free" carrier pair consisting of an electron and a hole.

An exciton may be annihilated either by additional thermal
"excitation, " i.e., by thermal dissociation accompanied by the
emission of an electron and a hole, or by the transfer of its energy

to the lattice.   An exciton may also be annihilated with the emission of a photon; this process is a special case of radiative recombination of nonequilibrium carriers, which will be discussed in Chapt. IV.

The most direct confirmation of the existence of excitons formed on the absorption of light outside the limits of the fundamental band is given by experiments in which exciton migration is observed over distances many orders of magnitude greater than the lattice constant of a crystal.   Apker and Taft [32] showed that in alkali-halide crystals with F-centers excitons migrated over distances not less than 1000 lattice constants from the places where they were generated.

Further confirmation of the existence of excitons in cuprous oxide (Cu_2O) and in CdS crystals was provided by the work of E. F. Gross and his co-workers [33], who found in these semiconductors a converging series of spectral lines, lying close to the fundamental band edge and similar to the spectral series of the hydrogen atom.[†]   The strong intensity of these absorption lines suggests that excitons are related to the host lattice and not to the impurities.   The exciton absorption line frequencies – for example, those of the "yellow" series of Cu_2O – are described by the formula

$$\nu_n = \nu_\infty - \frac{R_{ex}}{n^2},$$ (1.23)

where the quantum number is n = 2, 3, 4, 5, 6, 7, 8, 9, 10, ...; and ν_∞ is the frequency corresponding to the limit of the series, i.e., to the photodissociation of an exciton. R_{ex} is Rydberg's constant for an exciton, which is related to the Rydberg constant R for an atom by the expression

$$R_{ex} = \frac{R}{n_0^4} \frac{\mu^*}{m_n},$$ (1.24)

where n_0 is the refractive index, and the effective exciton mass μ^* is expressed in terms of the effective masses of electrons and holes in a crystal:

[†] The existence of similar absorption spectra has been confirmed also for CdSe, ZnS, HgI_2, PbI_2 and other semiconductors.

$$\frac{1}{\mu^*} = \frac{1}{m_n^*} + \frac{1}{m_p^*}. \tag{1.25}$$

Using the Wannier-Mott expression for the "diameter" of a weakly bound hydrogen-like exciton:

$$d_{ex} = 2\left(\frac{\hbar^2}{\mu^* q^2}\right) n_0^2 n^2, \tag{1.26}$$

E. F. Gross obtained values which were many times greater than the lattice constants

for $n = 2$ $\quad d_{2ex} \simeq 100\text{A},$
for $n = 10$ $\quad d_{10ex} \simeq 2500\text{A}.$

One of the consequences of the weak binding of an electron and a hole in an exciton is the dissociation of excitons in relatively weak electric fields. Beginning with a field intensity close to 6 kV/cm, the higher terms of the spectral series gradually disappear, and in a field of 30 kV/cm the whole "yellow" exciton series of Cu_2O disappears. Another interpretation of this interesting phenomenon is that the dominant effect is the spreading or shift of the fundamental band edge when an electric field is applied, as predicted by L. V. Keldysh and W. Franz [22, 23].

A careful investigation of the absorption edge of very pure single crystals of germanium and silicon, together with an analysis of the form of the fundamental band using the theory of vertical and nonvertical electron transitions, allowed MacFarlane et al. [20] to detect the very weak structure of the spectrum due to the formation of excitons. The spectral series, similar to those found by Gross, were not found in these crystals, because of the very low exciton dissociation energies amounting to 0.0027 eV for Ge and 0.012 eV for Si. We may assume that deeper exciton states exist near the strong absorption band edge which is related to vertical transitions. However, the presence of quite intense absorption due to nonvertical transitions with phonon participation makes experimental study of deep exciton states very difficult.

§ 5.   Absorption of Light Involving
the Photoionization or Excitation of Impurities
and Structural Defects

A.   Centers with Shallow Energy Levels

From measurements of the temperature dependence of the electrical conductivity and Hall e.m.f., it is known that in semiconductors with a narrow forbidden band and high permittivity, chemical impurities form "shallow" energy levels of the donor or acceptor type. Typical examples are phosphorus (donor) and boron (acceptor) in silicon crystals. It was shown some time ago that many elements of group V form donor centers in silicon which have approximately the same (0.045-0.05 eV) thermal ionization energy; in germanium crystals, the ionization energy of similar shallow donor centers is even lower – close to 0.01 eV.

These experimental observations have been explained qualitatively by means of the "hydrogen-like" model of impurity centers, according to which the impurity atom (for example, an atom of group V) behaves like a   hydrogen atom in a medium having a permittivity ε. From this well-known analogy, which will be shown to be a rough approximation, it follows that an electron transition from the ground state in a donor center to the conduction band should correspond to an infrared absorption in the photoconductivity band. Moreover, there should be absorption corresponding to transitions from the ground state to the excited states. The corresponding levels of the donor should lie between the ground level and the conduction band, and the absorption bands should be on the long-wavelength side of the photoionization threshold of the impurity.

Impurity atoms with low ionization energies are almost all ionized by thermal excitation at room temperature. Therefore, the phenomena related to electron transitions from shallow impurity levels to the band or from the band to these levels (in particular, the selective absorption and photoconductivity) can be observed only at sufficiently low temperatures.

Detailed experimental data on the absorption by centers with shallow levels are at present available for silicon crystals. In the case of germanium, the shallow-level absorption bands lie in a region of the spectrum ($\lambda > 100 \ \mu$) which is difficult to investigate. Nevertheless, some data on the absorption in germanium

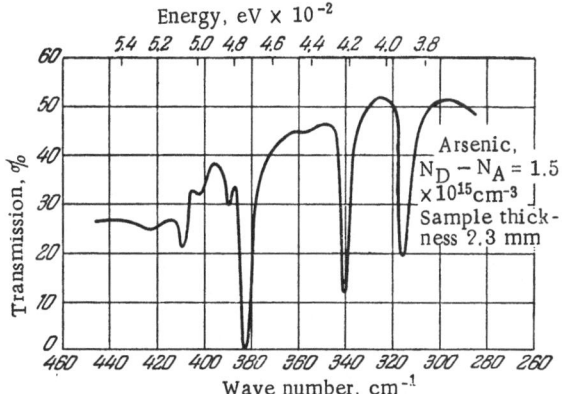

Fig. 15. Absorption spectrum related to photoionization and excitation of donor centers with shallow levels (arsenic in silicon)
[4]. $(N_D - N_A)$ is the difference between the concentrations of
shallow donors and acceptors.

in this region have been obtained by H. Y. Fan and his co-workers
[34]. We shall discuss in some detail the experimental data for
silicon and their theoretical interpretation.

The need to use low temperatures (T < 20°K) in these experiments is due not only to the low thermal ionization energy of the
centers but also in the fact that the absorption bands are broadened
by the interaction of electrons with lattice vibrations. Another
point which must be considered in investigations of the optical absorption is the interaction of impurity atoms with one another,
which already starts to broaden the absorption bands at impurity
concentrations of 2×10^{16} cm^{-3}. According to the approximate
theory of H. Y. Fan [1], the cross section σ' for the absorption
of a photon by an impurity center near the photoionization threshold, represented by the frequency ν_i, is given by the following
formula when $\nu > \nu_i$:

$$\sigma' = \frac{\alpha}{N} \approx \frac{5 \cdot 10^{-17}}{n} \frac{m}{m^*} \frac{E_H}{E_i} \left(\frac{\nu_i}{\nu} \right)^{3.5}. \qquad (1.27)$$

where α is the absorption coefficient, N is the impurity center concentration, m^* is the effective carrier mass, $n \approx \sqrt{\varepsilon}$ is the refractive index, $E_H = 13.6$ eV is the energy of ionization of the hydrogen

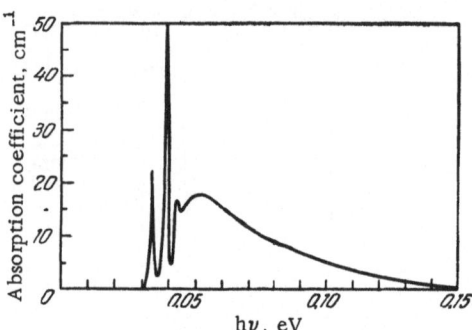

Fig. 16. Absorption spectrum related to boron atoms
in silicon.

atom, and $E_i = h\nu_i$ is the ionization energy of the impurity. The
accuracy of this expression is better than one order of magnitude.

An even rougher estimate, without allowing for the effective
mass, can be made using the expression for the absorption coefficient α_i

$$\alpha_{i\max} = \frac{N}{\varepsilon\left(\dfrac{E_i}{E_H}\right)} 10^{-17} \quad cm^{-1};$$ (1.28)

the above estimate must be used, for example, in the initial
studies of new semiconducting compounds when only the concentration of the main impurity and the values of ε and E_i are known.

A typical spectrum of the absorption by shallow donors (arsenic atoms) in silicon is shown in Fig. 15. The spectrum was
recorded at about 4.2°K. One of the bands lying near 316 cm^{-1}
was due to the presence of antimony. Similar spectra were obtained for bismuth, phosphorus, and antimony. The most important experimental facts are:

a. the presence of a clearly visible system of bands corresponding to excited states;

b. the similarity of the spectra (band systems) of various impurities;

c. the difference between the optical ionization energies of
various impurity centers (for example, P, As, Sb) in the same
host substance (silicon).

TABLE 4.   Experimental Values of the Ionization Energies of Impurity Centers with Shallow Levels in Silicon [4] (Energy Values in Thousandths of Electron-Volts)

Transitions	Acceptors			
	B	Al	Ca	In
$1s - 2p^1$	30.16 ± 0.12	54.91	58.26	142.06
$1s - 2p^2$	34.52 ± 0.12	58.57	61.80	145.73
$1s - 2p^3$	38.40	64.08	67.15	149.75
$1s - 2p^4$	39.65	64.99	68.26	150.88
$1s -$	41.46	66.90	70.68	153.36
$1s -$	42.13	67.55	71.36 ± 0.15	153.96
$1s -$	42.74	68.51 ± 0.12	72.28 ± 0.15	155.41 ± 0.2
$1s -$	43.85			

Transitions	Donors			
	P	As	Sb	Bi
$1s - 2p$, m = 0	34.5	42.27	31.8	59.50
$1s - 2s$		44.66		62.19
$1s - 2p$, m = ±1	39.5	47.47	36.5	64.57
$1s - 3p$, m = 0		48.35		65.47
$1s -$		49.87		67.13
$1s - 3p$, m = ±1	42.6	50.72	39.9	68.01
$1s -$	44.6	52.47 ± 0.3		69.12 ± 0.12

The acceptor centers (elements of group III in Ge and Si) behave similarly. Figure 16 shows a characteristic absorption spectrum due to a boron impurity in a silicon crystal.

Table 4 shows that the energies of transitions to excited states and the photoionization energy for shallow donor and acceptor levels can be determined from the positions of the absorption bands with great accuracy, which exceeds considerably the accuracy of the determination of the thermal ionization energy.

All the shallow donor centers have the same system of absorption bands and the same is true of the shallow acceptors but there is no direct correlation between the donor and acceptor bands. This indicates that the simple hydrogen-like model is imperfect because it predicts the same system of levels of acceptors and donors, the only difference being that of the effective mass values:

$$E_n = \frac{1}{n^2} \frac{q^2}{2\hbar^2} \frac{m^*}{\varepsilon^2}; \qquad n = 1, 2, 3, \ldots \qquad (1.29)$$

Moreover, no adjustment of the effective mass values for donors
or acceptors gives agreement with experiment.

The use of the hydrogen-like model for qualitative work is
justified by the fact that in the case of a hydrogen-like center the
effective radius of the Bohr orbit a^* is directly proportional to ε:

$$a^* = \frac{h^2}{q} \frac{\varepsilon}{m^*}. \qquad (1.30)$$

For silicon with $\varepsilon \approx 12$ and $m^* \approx 0.3\, m_e$, we obtain $a^* \approx 20$ A.
The lattice constant of Si is 5.42 A and therefore an electron
bound to an impurity atom is close to this atom only for a short
time, being "spread" over many crystal unit cells during the remaining time. It follows that the unknown potential distribution
near the atom does not greatly affect the motion of the electron.
These ideas apply even more to germanium crystals whose shallow donor and acceptor ionization energies are close to 0.01 eV
and whose $\varepsilon = 16$.

The principal assumption of the effective-mass theory (developed mainly by C. Kittel and A. Mitchell [35], as well as by
W. Kohn and J. Luttinger [36]), which is used to describe a system of shallow levels of impurity centers, is the postulate that
the form of the potential well near a nucleus of an impurity atom
is given by the equation

$$U = -\frac{q^2}{\varepsilon r}, \qquad (1.31)$$

where r is the distance from the atom. Naturally, the lower the
probability of finding an electron in the immediate vicinity of an
impurity atom, where the true distribution of the potential is unknown, the better should be the agreement with experiment. Apart
from the principal assumption mentioned above, the theory allows
for the band structure of the crystal in which the impurity atom is
located. The donor state is represented by a wave packet of Bloch
functions near the bottom of the conduction band. Thus, the kinetic
energy of an electron is governed by the energy of a minimum (or
minima) of the conduction band of a given semiconductor. Consequently, the systems of donor energy levels are different in different substances and the spectrum of absorption by donor centers may be considerably different from that of acceptors in the
same substance.

However, the effective-mass theory does not allow for the difference between different donor or acceptor atoms in the same crystal, which can be very considerable (Table 4). The theory of Kittel et al., which is a generalization of the idea of hydrogen-like centers, predicts – for shallow levels in silicon – an ionization energy E_{0i} equal to 0.029 eV, while the experimental values range from 0.039 eV (phosphorus) to 0.069 eV (bismuth). These differences are first of all due to the absence of any allowance for the true form of the potential curve near a center, and to the use of the average value of ε. Nevertheless, the theory allows us to identify several absorption bands connected with excited states and to calculate the difference between the energies of these states. For p-states, the calculated values are in good agreement with experiment. Therefore, we can find the quantity of greatest interest to us – the optical ionization energy – somewhat more accurately than by direct measurement, in which the edge of the band is usually difficult to determine.

A detailed account of the current state of the theory of impurity centers with shallow levels is given in Kohn's review [37].

B. Infrared Absorption Related to the Deep Levels of Impurities and Defects

Levels are referred to as deep when the probability of their thermal ionization at room temperature is small.

The existence of impurity centers or structural defects with deep levels will later be shown to govern such important physical properties of semiconductors as the recombination velocity of nonequilibrium carriers, the photoconductivity spectrum, and the luminescence spectrum.

It is natural to expect that beyond the long-wavelength edge of the fundamental absorption band, there should be absorption due to the photoionization or excitation of deep levels. Such absorption is, in fact, observed in semiconductors with a broad forbidden band (CdS, ZnS), as well as in silicon crystals in which structural defects are produced by irradiation with fast electrons or neutrons (cf. Chapt. V). Deep levels frequently correspond to secondary (etc.) ionization of an impurity atom or defect. In the semiconductors which have been studied most (germanium and silicon), the energy of ionization of the majority of deep impurity levels of Au, Fe, Co and other elements, have been determined

not from the optical absorption data, but by electrical measurements or from the spectra of the impurity photoconductivity. The
difficulties arising in the investigation of absorption are due to
the fact that impurities which replace atoms of the host lattice
(for example, Au in Ge or in Si) cannot be introduced in concentrations higher than $1\text{-}2 \times 10^{15}$ cm^{-3}. The impurity excess, introduced into the melt during crystal growth, is precipitated in the
form of colloidal occlusions.

Therefore, the interesting problems of the existence of excited states of impurity centers with deep levels have not yet been
investigated, even for Ge and Si.

The results of recent investigations of the infrared absorption in silicon containing radiation-induced structural defects indicate the existence of excited states [27, 28, 38]. However, a
quantitative interpretation of the available data is difficult because
there is as yet no theory of centers with deep levels. In such centers, an electron is strongly bound to an impurity atom (or, for
example, an interstitial atom), and because the spatial distribution of the wave function varies strongly within one lattice spacing,
we cannot use the effective-mass method. The bound electron
spends a large part of its lifetime in the immediate vicinity of the
center where the form of the potential curve is unknown.

Data on the optical ionization energies of many impurities
which form deep levels in germanium are listed in the work of
Newman and Tyler [39] and in Geballe's review published in [4],
which also includes data for silicon.

§6.   Infrared Absorption by Charge Carriers

A. Nonselective Absorption*

Nonselective absorption, which increases smoothly with wavelength up to very long wavelengths (more than 100 μ) is observed
beyond the fundamental absorption edge in semiconductors with
sufficiently high carrier densities. The absorption coefficient at
a given wavelength is then found to be approximately proportional
to the majority-carrier density, up to high impurity-center concentrations, and beyond that the absorption rises faster with the
density.

* Some authors (for example, Moss) call this type of absorption "metallic."

Nonselective absorption by carriers, which has been observed clearly in a wide range of wavelengths in n-type germanium [1], silicon [1, 40], indium antimonide, and in other semiconductors, is due to electron transitions within one band, for example, the conduction band. Quite frequently, this absorption is called, not very accurately, the "free-carrier absorption." In fact, such transitions are forbidden by the law of momentum conservation (the selection rule $k_i \approx k_f$) for carriers in a perfect periodic lattice not perturbed by thermal motion.

However, carriers can undergo transitions inside a band, because of thermal motion and structural defects, with which they may interact. The problem of the absorption of light by conduction electrons, which have a relaxation time θ, was discussed theoretically by Kronig [41]. Assuming formally that the "resonance" frequency of free electrons is equal to zero and introducing a resonance width governed by "damping" (the quantity θ was used as the parameter representing this damping), Kronig obtained a formula for the absorption coefficient, similar to the classical expression of Drude for an electron oscillator:

$$\alpha = \frac{4\pi}{cn} \frac{n_e q^2}{m^*} \frac{\theta}{1 + \omega^2 \theta^2} , \qquad (1.32)$$

where n_e is the electron density in the conduction band, c is the velocity of light, n is the refractive index, and $\omega = 2\pi \nu$ is the angular frequency. In semiconductors with high mobility, the condition $\omega^2 \theta^2 \gg 1$ is usually satisfied. The relaxation time θ can be estimated knowing the electron mobility μ,

$$\theta = \frac{\mu m^*}{q} , \qquad (1.33)$$

and hence it follows that the coefficient of absorption by charge carriers is

$$\alpha = \frac{n_e q^3}{c n \pi \mu m^*} \frac{1}{\nu^2} . \qquad (1.34)$$

Thus, the absorption by carriers should be characterized by a dependence of the type

$$\alpha \sim \lambda^2. \qquad (1.35)$$

The absorption observed in n-type germanium at room temperature was compared by Kane [42] with the expression in Eq. (1.34).
He found that the best agreement with experiment was obtained if
the mean effective electron mass m_e^* was $0.11-0.22\,m_e$, expressed as follows

$$\frac{1}{m_e^*} = \frac{1}{3}\left(\frac{1}{m_{e\parallel}^*} + \frac{2}{m_{e\perp}^*}\right),$$ (1.36)

where $m_{e\parallel}^*$ is the "longitudinal" and $m_{e\perp}^*$ the "transverse" effective mass of a free electron. The average effective mass deter-
mined by the cyclotron resonance method is $0.12\,m_e$ for Ge.

Later, the theory of absorption by carriers was refined by
Fan, who deduced expressions for the absorption coefficient α,
allowing both for the interaction of electrons in the conduction
band (or holes in the valence band) with the lattice, and for the
influence of impurities and defects [1]. According to Fan, when
carriers are scattered mainly by the acoustical modes of lattice
vibrations:

$$\alpha = \frac{4\pi\sigma}{cn} = \frac{4\pi}{cn}\,\frac{4}{9\pi^{1/2}}\,\frac{\sigma_0}{0^2\omega^2}\left(\frac{\hbar\omega}{kT}\right)^{1/2}\left\{\left(1+\frac{2E_k}{\hbar\omega}\right)\left(1+\frac{E_k}{\hbar\omega}\right)\right\},$$ (1.37)†

where σ_0 is the conductivity. The braces $\{\ \}$ denote averaging
over the initial kinetic energy of electrons (or holes) E_k. In the
classical statistics case, $E_k \approx kT$. Therefore, in contrast to
the classical theory of Kronig, we have, from the last expression,

$$\alpha \sim \lambda^{3/2}.$$ (1.38)

At very high concentrations of ionized impurity centers, the absorption coefficient α should vary with wavelength as $\lambda^{7/2}$; at
any given wavelength, the absorption coefficient should be proportional to the square of the impurity-center concentration [43].

The absorption by free carriers in p- and n-type silicon was
investigated by the present author [40] at wavelengths from 1.1
to 11 μ. In p-type crystals containing at least 5×10^{15} cm^{-3} holes,
the absorption coefficient increased smoothly with wavelength

---

† In deducing Eq. (1.37), Fan did not allow for induced emission, and this led to the
appearance of an additional multiplier $[1 - \exp(-\hbar\omega/kT)]$.

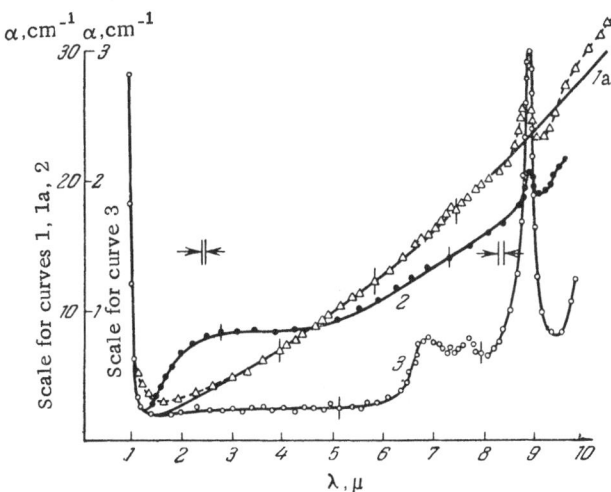

Fig. 17. Absorption spectrum of p- and n-type silicon single crystals at 300°K. 1) p-Type silicon, $p_0 = 1.5 \times 10^{17}$ cm^{-3}; 1a) theoretical dependence of α; 2) n-type silicon, $n_0 = 4 \times 10^{17}$ cm^{-3}; 3) p-type silicon, $p_0 = 1.5 \times 10^{15}$ cm^{-3}

(Fig. 17); for hole densities up to 3×10^{17} cm^{-3}, the dependence $\alpha = f(\lambda)$ was very close to that predicted by the theory of Fan. The continuous curve 1a in Fig. 17 shows the theoretical dependence. The absolute values of the absorption coefficients in the spectral region for which the condition $\alpha \approx \lambda^{3/2}$ was satisfied, were used to calculated the effective hole mass m_n^* of silicon. According to the present author [40], $m_p^* = (3.5 \pm 0.3) \times 10^{-28}$ g, i.e., $m_p^*/m_e = 0.38$, which is in good agreement with the value $0.39 m_e$ obtained by Dexter and Lax by the cyclotron resonance method [44]. In p-type silicon crystals with relatively low hole densities (less than 5×10^{15} cm^{-3}), the present author [40] found strong deviations from the $\lambda^{3/2}$ law (Fig. 17, curve 3), which appeared in the form of an "absorption" which depended weakly on λ in the region 1.1-7.0 μ. Control tests have shown that the observed effect is a volume one but it has not been established whether the attenuation of the transmitted beam is due to the intrinsic absorption of light or to the scattering of radiation on defects, stresses, and impurity aggregates [45].

In contrast to p-type silicon, none of the investigated n-type crystals [40] exhibited a dependence of the $\alpha \sim \lambda^{3/2}$ type. A typical

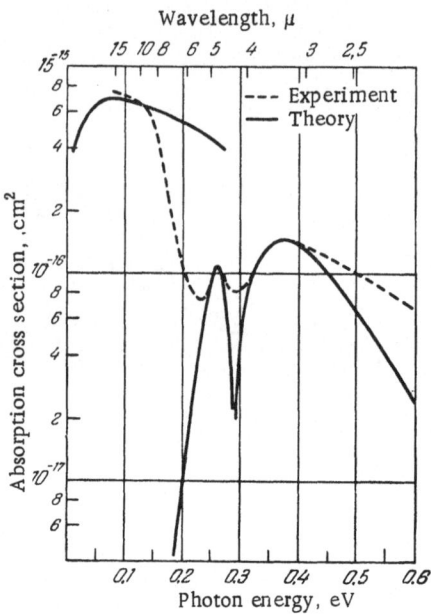

Fig. 18. Absorption by holes, injected into n-type germanium [17].

dependence on wavelength for an n-type crystal with a high elec-
tron density ($n_e \approx 4 \times 10^{17}$ cm^{-3}) is also given in Fig. 17 (curve
2). This curve has an inflection in the region of 2.5-3 μ, after
which α increases with λ more slowly than $\lambda^{3/2}$. We cannot re-
ject the possibility that the observed wavelength dependence is
due to the superposition of a spectrum of the $\alpha \sim \lambda^{3/2}$ type and a
wide band with a maximum near 2.5 μ, as assumed by Fan [17].
According to the investigations of V. A. Yakovlev [46], allowance
for scattering by optical phonons and for the complex band structure of silicon, not made in Fan's theory, leads us to the conclusion that the "maximum" is also due to the absorption by conduction electrons.

The influence of the nonselective absorption by carriers
makes it very difficult to detect weak absorption bands related
to lattice vibrations and impurities. In view of this (particularly
in studies of the infrared spectra of new semiconducting materials), it is very important to lower the carrier density in some
way (for example, by introducing compensating impurities or by
lowering the test temperature).

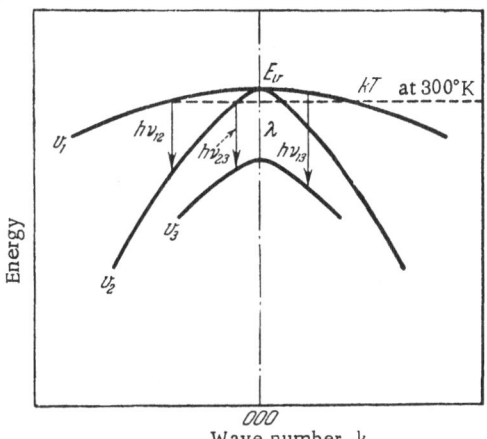

Fig. 19. Complex structure of the valence band of
germanium [42].

B. Selective Absorption

In contrast to the absorption characterized by a monotonic increase in α with wavelength, under certain circumstances there
may exist relatively narrow spectral absorption bands also related
to the interaction of radiation with carriers. The best known case
of selective absorption by carriers is the absorption by holes in
germanium crystals. Very soon after the first successful growing
of sufficiently large single crystals of n-type Ge, it was established that in the region beyond the fundamental band edge, near
3.4 and 4.7 μ, and at wavelengths greater than 10 μ, there were
absorption bands whose intensities were proportional to the hole
density. It was possible to demonstrate that an exactly similar
absorption spectrum should be observed in n-type germanium crystals under "nonequilibrium" conditions, for example, when holes
are injected into an n-type region by passing direct current through
a crystal with a p-n junction [47]. The spectrum of absorption by
holes in a germanium crystal is shown in Fig. 18, where the ordinate gives the absorption cross section, i.e., the absorption coefficient divided by the hole density in the test crystal.

Briggs and Fletcher [47] have interpreted the observed absorption spectrum using the information on the complex structure
of the valence band of germanium; this band has three "branches,"

two of which (v_1 and v_2) coincide near the point k = 0, and the third (v_3) lying below the other two. The absorption band at λ = 3.4 μ (0.37 eV) is ascribed to the electron transition $v_3 \rightarrow v_1$; the band at λ = 4.7 μ (0.27 eV) is ascribed to the transition $v_3 \rightarrow v_2$; and, finally, the band at λ = 15 μ (0.08 eV), the shape of which is least known, is ascribed to the transition $v_2 \rightarrow v_1$ (cf. Fig. 19). It should be noted that the separation of the lowest band v_3 from the bands v_1 and v_2 has been determined from the position of the absorption band at λ = 4.7 μ, since the data obtained by other methods simply show that such separation should exist but cannot be used to determine it. The theory of transitions between the "branches" of the valence band of Ge was developed by Kane [42].

The strong absorption by holes in germanium has been utilized in an experimental investigation of the spatial separation of injected holes, for determining the diffusion length [48], and for the modulation of infrared radiation [49, 50].

In contrast to germanium crystals, and in spite of the great similarity of the valence-band structure, silicon crystals exhibit practically no selective absorption by holes. Investigation of the infrared absorption spectra of certain intermetallic compounds, for example, GaSb and InAs, has revealed the presence of selective absorption by carriers, which – as in germanium – is related to transitions between the "branches" of a complex valence band [51, 52].

§7. Influence of Electrically Inactive Impurities on the Infrared Absorption in Semiconductors

The presence of impurities in semiconducting materials, in particular in germanium and silicon, has been usually deduced from measurements of the electrical conductivity, the Hall effect, and the lifetime of minority carriers. Relatively recently, it has been found that, very often germanium and silicon single crystals contain very considerable amounts of oxygen or hydrogen, in addition to the electrically active impurities, which can capture carriers or supply them to the respective band by ionization. The concentration of such impurities, in particular oxygen, sometimes exceeds 10^{18} cm^{-3}. Oxygen atoms may exist in silicon crystals in two states. In one of these, usually in melt-grown single crystals, the oxygen atoms do not affect the electrical properties of

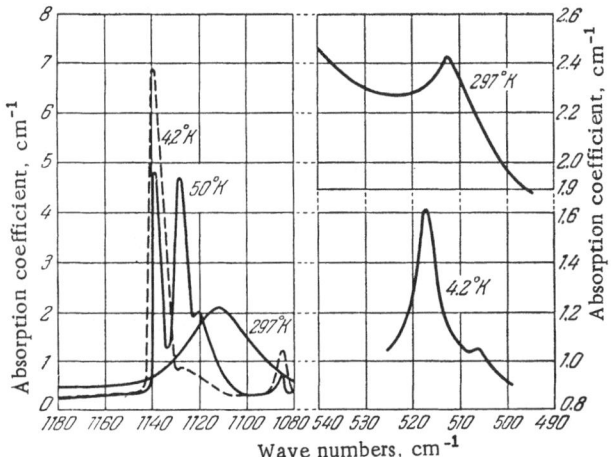

Fig. 20. Absorption bands due to the excitation of vibrations
of oxygen atoms in silicon. The oxygen contained 12% O^{18} [53].

the silicon. However, subsequent heat treatment may alter quite
markedly the electrical conductivity and other properties of oxy-
gen-bearing silicon.

   The question of the existence of dissolved oxygen in silicon
and of its influence on the properties of the latter has been an-
swered by the discovery of characteristic infrared absorption
bands which are due to the vibrations of oxygen atoms and neighboring silicon atoms. Since the investigation of the "electrically
inactive" impurities in semiconductors is of great practical importance but has not been intensively studied, we shall consider
the case of oxygen in silicon in some detail.

   The strongest absorption band, due to the vibrations of oxygen
bound to silicon has a wave number which is close to 1106 cm^{-1}
($\lambda = 9.1 \mu$) at 300°K. Another band is situated at longer wavelengths, close to 515 cm^{-1}, and third, the weakest, is located near
1205 cm^{-1}. Experiments on silicon containing oxygen enriched
with O^{18} have shown that there is a pronounced isotopic splitting
and, consequently, the observed absorption bands are due directly
to oxygen atoms. Figure 20 shows these absorption bands at various temperatures for a sample of silicon containing oxygen   in
which 12% of the oxygen atoms are O^{18}. Although silicon dioxide
(SiO$_2$) has a strong absorption band near 1100 cm^{-1}, an investigation of the isotopic splitting and especially of the band intensity

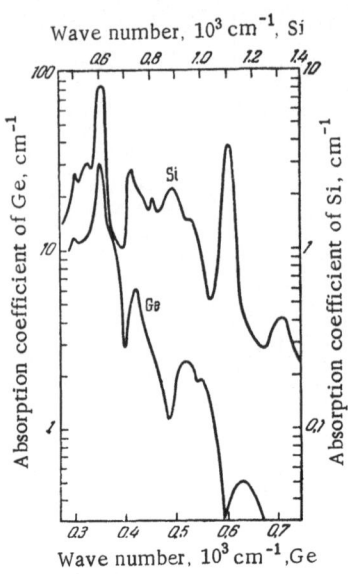

Wave number, 10^3 cm^{-1}, Si

Wave number, 10^3 cm^{-1}, Ge

Fig. 21. Absorption bands due to the excitation of crystal lattice vibrations in Ge and Si (with the exception of the 1106 cm^{-1} band of Si, which is due to oxygen).

has indicated that the "vibrator" responsible for the absorption includes only one atom of oxygen, i.e., the vibrations seem to occur in an Si_2O "molecule" [53, 54]. A detailed investigation of the temperature dependence of the absorption bands associated with oxygen in silicon has provided considerable information on the structure, i.e., the geometry and binding forces, of the vibrating Si_2O "molecule." At present, infrared spectroscopy is being used as a simple method of quantitative analysis of the initial content of oxygen in silicon, beginning with concentrations of 5×10^{15} cm^{-3}. Heat treatment at temperatures above 1000°C, as well as bombardment with neutrons [28, 38], reduces to absorption intensity. Kaiser [55] has shown that in the former case oxygen atoms are "condensed" and SiO_2 is formed. When the concentration of silicon dioxide is sufficiently high, its presence can be detected by the resultant Rayleigh scattering of light.

§ 8.   Absorption of Light Involving Excitation of Crystal-Lattice Vibrations

Like the excitation of vibrations in which impurity atoms participate, the absorption of light which excites vibrations of the host crystal lattice is not accompanied by photoionization. This type of absorption is very typical of ionic crystals, all of which have very intense absorption bands in the far infrared and a reflectivity maxima (reststrahlen) slightly displaced from the band [5, 56]. The problem of the vibration spectrum of a crystal lattice and of the frequency at which the absorption maximum should occur has been discussed in Kittel's book [57].

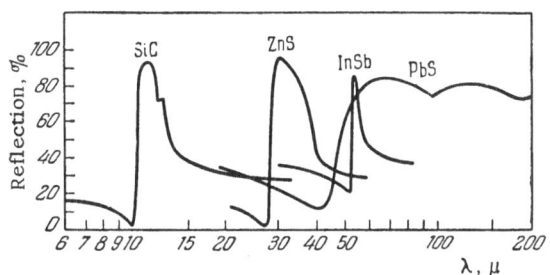

Fig. 22. Reflection maxima of semiconducting compounds,
due to the excitation of crystal lattice vibrations.

The absorption or reflection intensity in the "reststrahlen"
band decreases with the reduction in the ionicity of the binding,
i.e., with the effective charge of atoms in a crystal. Thus, covalent crystals of group IV elements – diamond, silicon, and germanium – should not, according to the simple theory, have absorption bands associated with the excitation of lattice vibrations,
since the effective atomic charge (which represents the degree of
ionicity of the bonds) is zero for these crystals. However, experimental studies of them have indicated the presence in each
of these crystals of a system of absorption bands which are undoubtedly due to lattice vibrations (cf. Fig. 21). The absorption
and intensity of these bands have been found to be independent of
the type and concentration of the electrically active impurities.
Collins and Fan [58] have established that the intensity of absorption in these bands is proportional to the rms displacement of
atoms under thermal excitation. To explain the existence of absorption bands in the case of group IV elements, it is necessary
to make the additional assumptions that the thermal motion of
atoms or the presence of structural defects deforms the charge
distribution in the crystal, producing electric dipoles with which
the incident infrared radiation may interact.

Lax and Burstein have shown [59] that an electric moment of
the second order of smallness appears if charges induced by vibrations of one type oscillate under the action of vibrations of another
type. Such a process is possible if two phonons are created simultaneously or if one phonon is created simultaneously with the
annihilation of another. Thus, the infrared absorption bands associated with the lattice vibrations of covalent crystals are com-

bination bands associated with two phonons having the same wave vector but belonging to different branches (modes) of the lattice vibration spectrum. A theoretical calculation of the absorption spectra of the lattice vibrations of germanium and silicon does not agree well with the experimental results.

In contrast to true covalent crystals, semiconducting compounds very similar to these crystals (for example, silicon carbide or indium antimonide) exhibit strong absorption bands and associated reflectivity maxima (cf. Fig. 22). The absorption band of InSb is much narrower than the corresponding bands of compounds with stronger ionicity of bonds, for example, ZnS or PbS, which indicates that the effective charge of atoms in InSb is small. This is due to the small difference between the electronegativities of indium and antimony. Silicon carbide, on the other hand, has a strongly polar binding, which is in agreement with the large difference of the electronegativities of Si and C.

Chapter II

PHOTOIONIZATION AND PHOTOCONDUCTIVITY IN SEMICONDUCTORS

Photoconductivity, i.e., the change in the electrical conductivity of a crystal under the action of absorbed radiation, is being increasingly utilized to study other processes in semiconductors. In the semiconductors studied most – germanium, silicon, and certain intermetallic compounds – it has become possible to control, within a wide range, the photosensitivity spectrum, response, and other properties important in the practical applications of these materials.

Photoionization is the primary process which initiates photoconductivity. Much of the work on photoconductivity has practically ignored the problem of photoionization, therefore we shall use new results to consider the photoionization in germanium and silicon in greater detail than the process of recombination, which is dealt with thoroughly in Ryvkin's review [1]. The processes . of recombination have also been analyzed in a monograph by R. Bube [2].

§9. Principal Quantities and Relationships in Photoconductivity

The electrical conductivity of a homogeneous semiconductor σ is given by the formula

$$\sigma = q\,(n\mu_n + p\mu_p), \qquad (2.1)$$

where q is the electronic charge, n and p are the free electron and hole densities, μ_n and μ_p are their mobilities. A photoconductivity $\Delta\sigma$ appears if, as a result of the absorption of radiation, the values of n and p increase compared with their values at thermal equilibrium[†]

$$\Delta\sigma = q\,(\Delta n\,\mu_n + \Delta p\,\mu_p). \qquad (2.2)$$

In semiconductors with a wide forbidden band or at low temperatures, the values of Δn and Δp may be considerably higher than the corresponding equilibrium (dark) densities n_0 and p_0. In semiconductors with a high equilibrium carrier density (due to a narrow forbidden band or the presence of impurities), the effect of radiation is usually limited to only small deviations from the dark electrical conductivity. In view of this, completely different approaches must be used for the photoelectric effects in semiconductors which can be classed almost as dielectrics and for these effects in "normal" semiconductors.

In an inhomogeneous semiconductor, for which the values of n_0 and p_0 vary from one region to another, the photoconductivity may be due to more complex effects, such as a change in the resistance of barriers between grains (crystallites) as a result of radiation. If the grains in a polycrystalline semiconductor sample are separated by layers of much higher electrical resistivity, the current through the sample will be governed mainly by the properties of these layers and not by the electrical conductivity of the grains themselves. A change in the carrier density in the region of such a layer (due to the absorption of radiation) may strongly reduce the resistivity, i.e., it may manifest itself as photoconductivity.

In describing the photoconductivity in inhomogeneous semiconductors, one sometimes uses the concept of "effective mobility" μ^*. A material with barriers at grain boundaries is then regarded as homogeneous with a carrier density corresponding to the regions of higher conductivity, and it is assumed that the measured overall electrical conductivity corresponds to the mobility μ^* in some equivalent homogeneous material. Thus, in this case,

$$\Delta\sigma = q\,(n\,\Delta\mu_n^* + p\,\Delta\mu_p^*). \qquad (2.3)$$

[†] If we define the photoconductivity simply as an increase in n and p on the absorption of radiation, as is done by Bube [2], the effect of heating a semiconductor by the absorption of radiation should also be considered as photoconductivity, which seems incorrect.

This is an arbitrary description and does not reflect the essential features of physical processes occurring in a semiconductor. *

The principal expressions used in the analysis of steady-state photoconductivity, which are also definitions of the "average life-times" τ_n and τ_p of nonequilibrium carriers, are the formulas

$$\left.\begin{array}{l} g_n \tau_n = \Delta n, \\ g_p \tau_p = \Delta p. \end{array}\right\} \tag{2.4}$$

Under steady-state conditions, the excess electron density Δn (or hole density Δp) in the conduction (valence) band is equal to the rate of generation g, i.e., to the number of carriers generated by radiation in unit volume in 1 sec multiplied by the average life-time of these carriers. The photoconductivity is then given by

$$\Delta \sigma = g q \left(\mu_n \tau_n + \mu_p \tau_p \right) \tag{2.5}$$

if $g_p = g_n = g$. In general, the lifetimes τ_n and τ_p depend not only on the composition of the semiconductor, i.e., on the host sub-stance and the presence of defects and impurities, but also on tem-perature and on the intensity of radiation.

The quantity g, which is known as the generation function or the rate of generation, has the dimensions of $cm^{-3} sec^{-1}$. The value of g is proportional to the intensity of excitation (light or charged particles) and depends on the properties of the semicon-ductor.

In analyzing photoconductivity it is essential to distinguish the primary process of generation (i.e., the photoionization or, for example, ionization by fast particles). If the generation func-tion g is known, the description of subsequent processes of the motion and recombination of nonequilibrium electrons and holes is, in the majority of cases, the same for excitation by light as for excitation by gamma rays or charged particles. Therefore, in-stead of attempting to introduce new terms such as "γ-conduc-tivity," "cathode-ray conductivity," "e-conductivity," "α-con-ductivity," etc., it is more useful to employ Vul's concept [3] of "radiation conductivity," which unifies all these effects, or to use

---

* This formal description does not allow for effects which actually alter the carrier mobility during irradiation.

the older term "photoconductivity" to denote the general case of excitation of nonequilibrium carriers at the expense of radiation energy.

In connection with the difference in the approach to the effects associated with nonequilibrium electron processes in semiconductors with a considerable dark conductivity (for example, Ge and Si under normal conditions) and in semiconductors with a wide forbidden band (CdS, SiC, crystal phosphors), different authors have assigned different meanings to the concept of "carrier lifetime," which are listed below.

The free-carrier lifetime, * i.e., the average period during which an electron or hole participates in the transport of charge in space. Thus, for an electron, this is the time it spends in the conduction band, and for a hole it is the time it spends in the valence band. The values of τ_n and τ_p in Eq. (2.5) are, in fact, the free-carrier lifetimes. This lifetime is limited by recombination, and by the loss from the crystal to an electrode or the surface, if the conditions are such that the lost electron (hole) cannot be replaced by another. If the loss of an electron at one electrode is compensated by the injection of another at the opposite electrode, the existence of a free carrier does not end there because carriers are indistinguishable. In the presence of nonrecombination capture centers, usually known as traps or trapping levels, the time spent by carriers in a localized state at these traps is not included in the lifetimes τ_n and τ_p.

The excited-state lifetime includes the whole period during which a carrier exists: from the moment of excitation to recombination or complete loss from the crystal. It includes also the time spent by the carrier in a localized state (at traps). This lifetime may be considerably longer than the free-carrier lifetime.

The electron—hole pair lifetime is the lifetime which governs the so-called ambipolar diffusion and represents the lifetime of a pair of free carriers which are not bound into an exciton. If the electron or the hole is localized, the pair lifetime is reduced by the time spent by one of the carriers at a trapping center.

The lifetimes of majority and minority carriers are also widely used. If the free-carrier density is considerably greater than

---

* It is more correct to speak of "charge carriers" or "current carriers" than of "free carriers." The former two terms are used in Soviet literature. For brevity, we shall use the word "carriers" [12].

the concentration of recombination capture centers, which frequently happens in Ge, Si, and intermetallic compounds (for example, in InSb), the lifetimes of majority and minority carriers are equal.

If the density of free carriers is lower than the concentration of recombination centers (which is typical of CdS, ZnS, and similar compounds), the lifetime of majority carriers may be considerably longer than the minority-carrier lifetime. The simple definition of either of these lifetimes is valid only in the case of steady-state conditions ($\Delta\sigma$ = const) and in the case of exponential rise or decay of $\Delta\sigma$ with time.

The photosensitivity of a semiconductor acted on by radiation is its photoconductivity per unit excitation intensity, i.e., the change in the electrical conductivity $\Delta\sigma$ divided by the power of the exciting radiation absorbed in unit volume.

The specific photosensitivity S is sometimes defined as the quantity obtained by multiplying the photoconductivity $\Delta\sigma$ by the square of the distance l between the electrodes of a homogeneous semiconductor and dividing by the absorbed radiation power P:

$$S = \frac{\Delta\sigma l^2}{P}\left[\frac{cm^3}{ohm \cdot W}\right]. \tag{2.6}$$

If the photoconductivity is a linear function of P and of the applied voltage V, the quantity S is the property of a semiconducting sample.

The photocurrent ΔJ in a homogeneous semiconductor can be expressed as follows

$$\Delta J = qGA, \tag{2.7}$$

where G is the total generation equal to the product of g and the semiconductor volume for the case of weakly absorbed light; the quantity A, known as the "photoconductivity gain," is equal to the ratio of the free-carrier lifetime to the transit time t taken by a carrier to travel from one electrode to another in an electric field $\mathcal{E}$ = V/l , where V is the applied voltage and l is the length of the sample.

If the absorbed radiation generates electron and hole pairs,

$$A = \frac{\tau_n}{t_n} + \frac{\tau_p}{t_p}, \tag{2.8}$$

where t_n and t_p are, respectively, the transit times for electrons and holes. Each of these times is inversely proportional to the carrier mobility:

$$t = \frac{l^2}{\mu V}.$$ (2.9)

Thus,

$$A = (\tau_n \mu_n + \tau_p \mu_p) \frac{V}{l^2}$$

and

$$\Delta J = qG\, (\tau_n \mu_n + \tau_p \mu_p)\, \frac{V}{l^2}.$$ (2.10)

We must remember that in this case a carrier which has reached an electrode is replaced by a carrier of the same type injected from the opposite electrode. The expression for ΔJ can also be obtained directly from Eq. (2.5) by multiplying it by the field intensity.

The generation rates are governed by the photoionization quantum yield η, i.e., by the number of electron-hole pairs generated by the absorption of a photon from the fundamental band of wavelengths, or by the number of carriers of one type produced by the photoionization of impurities.

§10. Photoionization Quantum Yield and Its Spectral Dependence

A. Concept of the Quantum Yield and Its Determination from the Photocurrent in Homogeneous Crystals

Current theories of the internal photoeffect in semiconductors have begun to develop parallel with other branches of physics based on the quantum concept of the absorption and emission of light, after the classical experiments of Gudden and Pohl [4], and the work of Tartakovskii et al. [5]. They have shown that certain crystals (alkali-halide compounds, ZnS, and diamond) exhibit an internal photoeffect (similar to the external photoeffect) due to electronic and not photochemical processes. Selecting their experimental conditions correctly, the cited authors have shown that the so-called "primary" photocurrent is a result of internal photoionization with a quantum yield equal to unity. In the preceding section, we have shown that, because of the indistinguishability of carriers of a given type and because the lifetime τ may exceed the transit time t, it is possible to have photocurrents for which the ratio of the transported charge to the number of absorbed quanta differs very greatly from unity, i.e., the photoconductivity quantum yield may differ by a factor of A from the photoionization quantum yield.

This condition, as well as the frequently observed nonlinear dependence of the photocurrent on the excitation intensity and the slow response of photoconductivity, has made the interpretation of experiments difficult. Recently, the tendency has been to investigate the process of photoionization in quantitative terms [6, 7]. Experiments on single crystals of the well-known semiconductors (Ge, Si, InSb, etc.) have confirmed that photoionization is a typical quantum process, and have yielded important quantitative data on a new effect: the impact multiplication of photoelectrons and holes.

In view of the importance of the problems related to the photoionization quantum yield η, we shall consider in detail the methods used to measure its value.

These methods reduce to:

a. measuring photocurrents in homogeneous crystals;

b. measuring photocurrents in inhomogeneous crystals with p-n junctions.

The former method has been used by Gudden and Pohl, and its variants have been employed by many others [8, 9].

Early experiments, which have established the quantum nature of photoionization, have been described by Haynes and Hornbeck [45]. Gudden and Pohl have distinguished primary photocurrents, i.e., those associated directly with the absorption of light, and secondary photocurrents, the nature of which they have been unable to establish but have suggested that they may be due to changes produced in a crystal by the passage of the primary current. From the modern point of view, the division of photocurrents in semiconductors into primary and secondary is unnecessary since the description of photoconductivity given above can be used to explain the characteristics of the phenomenon in a natural way.

In designing their experiments, Gudden and other workers started from the analogy between an illuminated homogeneous crystal with electrodes and a gas ionization chamber. Their elec-

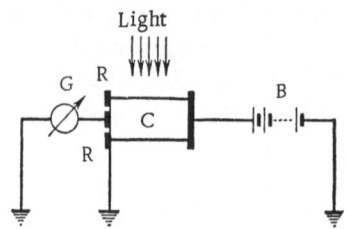

Fig. 23. Circuit used to measure
the photocurrent in crystals [45].
C is the crystal; G is a ballistic
galvanometer; B is a battery; R is
a guard ring for eliminating errors
due to surface conduction.

trodes had such properties that the photocarriers which had reached one of them ended their existence there without causing new carriers to enter the crystal, i.e., the lifetime was shorter than or equal to the transit time. Measurements were carried out using short excitation pulses and a ballistic galvanometer (Fig. 23). It was established that prolonged illumination of a crystal to which a field was applied produced polarization because of the capture of photocarriers by traps. To destroy this polarization, the crystals were heated or illuminated with long wavelengths after each photocurrent pulse; such treatment liberated the carriers from the traps and destroyed the polarization space charge.

To generate photocurrent pulses, monochromatic light of low intensity was used and the extent of the absorption of this light by the crystal was known.

It was found that under such experimental conditions:

a. the photocurrent was proportional to the excitation intensity;

b. in weak electric fields, the photocurrent was proportional to the field intensity $\mathscr{E}$; on further increase in $\mathscr{E}$ the current reached saturation;

c. the photocurrent rose or decayed practically instantaneously on the application of the excitation or at the end of it.

Moreover a quantitative analysis of the dependence of the photocurrent, or better of the transported charge ΔQ (which appeared on the illumination of a crystal and was measured with a ballistic galvanometer), on the field intensity $\mathscr{E}$ in the sample showed that

$$\Delta Q = G^\beta q \, \frac{x_+ + x_-}{l}, \qquad (2.11)$$

where x_- and x_+ are the so-called "drift lengths" or "displacements" of an electron and a hole, i.e., the distances travelled by photocarriers along the field direction up to the moment of capture; l is the length of the crystal; and θ is the duration of the light pulse. Since the maximum value of the sum of the drift lengths is $x_+ + x_- = l$, Eq. (2.11) shows that the number of electron charges transported from one electrode to the other cannot exceed unity in the case when one electron and one hole are formed on the absorption of one photon. In accordance with the formulas (2.10)

$$x_+ = \mu_p \mathscr{E} \tau_p, \quad x_- = \mu_n \mathscr{E} \tau_n. \tag{2.12}$$

The primary photocurrents have been distinguished and investigated in diamond, zinc sulfide, alkali-halide crystals with defects (F-centers), and in certain other substances [10]. The main conclusion of these experiments was that close to the edge but outside the absorption band, the photoionization quantum yield $\eta = \Delta Q / q N_{h\nu}$ for saturation photocurrents was very close to unity; here, $N_{h\nu}$ is the number of absorbed quanta. Within the absorption band, i.e., when the absorption coefficient was higher and the photoionization process concentrated in the surface layer of a crystal, the ratio $\Delta Q / q N_{h\nu}$ always fell. We shall show later that this was due to the very rapid recombination at the crystal surface. The value of η is called the "quantum yield" in analogy with the quantum yield for the external photoeffect.

Measurements of the photocurrents in homogeneous crystals for the purpose of determining the photoionization quantum yield of typical semiconductors with a narrow forbidden band (cuprous oxide, Cu_2O; germanium; indium antimonide) were carried out by Ryvkin [8], Goucher [9], and Tauc [11]. In contrast to the experiments on "primary" photocurrents, the establishment of strong fields in semiconductors at room temperature may give rise to strong heating of the samples by the Joule effect; on the other hand, at low temperatures, even in relatively weak fields, impact ionization of impurities occurs in semiconductors which, in the case of germanium, again leads to a strong rise in the free-carrier density.

Since it is impossible to use strong electric fields and thus reach photocurrent saturation, it is necessary to use independently determined values of the carrier mobility and lifetime. At present, the methods of determining the carrier lifetime in semiconductors are well developed and Ryvkin [8], Goucher [9], and Tauc [11] were able to deduce from their experiments reliable values of the photoionization quantum yield, which was, as expected, very close to unity near the fundamental band edge.

B. Determination of the Photoionization Quantum Yield in Semiconducting Crystals with p-n Junctions

Semiconductor crystals with deliberately produced p-n junctions can be used to detect photoelectric effects without applying external fields. In contrast to the experiments on homogeneous samples, the photo-emf or photocurrent in crystals with junctions can be measured in such a way as to eliminate the phenomena connected with the capture of minority carriers by traps.

If one region of a semiconductor crystal is made p-type by introducing an acceptor impurity and a neighboring region is made n-type by introducing donors, a potential barrier appears at the boundary between these two regions. It is convenient to represent this barrier as the result of the contact of two crystals, p- and n-type, which were initially separate. When these crystals are joined, some electrons cross over from the n-type into the p-type region because of the existence of carrier density gradients; holes cross over in the opposite direction. This leakage of carriers continues until a sufficient space charge is accumulated near the junction (positive charge in the n-type region and negative in the p-type). This space charge gives rise to a contact potential difference which should produce drift currents opposite to diffusion currents. The height of the potential barrier V_k when the diffusion currents are balanced by the conduction currents, i.e., at equilibrium, is given by the formula

$$V_k = \frac{kT}{q} \ln \frac{n_p}{n_n} = \frac{kT}{q} \ln \frac{p_n}{p_p} = \frac{kT}{q} \ln \frac{n_n p_p}{n_i^2}, \qquad (2.13)$$

where n_p and p_n are the equilibrium densities of minority carriers; n_n and p_p are the densities of majority carriers; n_i is the density of carriers of any type in an intrinsic semiconductor at a temperature T (Fig. 24).

If the radiation energy is absorbed and nonequilibrium carrier pairs are generated near a p-n junction, the holes which have reached the junction as a result of diffusion in the n-type region will be driven by the electric field into the p-type region; similarly,

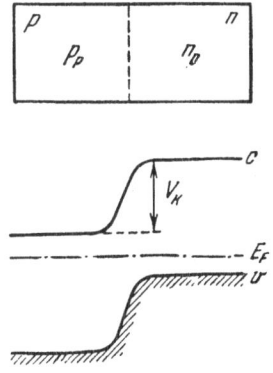

Fig. 24. Band structure of a semiconductor with a p-n junction: c is the conduction band; v is the valence band; E_F is the Fermi level; p_p and n_n are the equilibrium majority carrier densities; V_k is the contact potential difference.

the nonequilibrium electrons will cross over into the n-type region, i.e., the following "generation" current will flow through the junction

$$J = \beta G q = J_p + J_n, \qquad (2.14)$$

where G is the total number of pairs generated by radiation in 1 sec, and β is a dimensionless quantum efficiency* which depends on the sample geometry, velocities of volume and surface recombination, and on ambipolar diffusion coefficients in the interior of the semiconductor.

The value of the quantum efficiency in the one-dimensional case was discussed by Pfann and van Roosbroeck [13], who stated the problem, established the boundary conditions, and reported the final expression for β without giving the intermediate steps. We shall state here the one-dimensional problem and the final formulas, which we shall use later in connection with the determination of the photoionization quantum yield.

We shall consider a semiconductor bounded by the plane x = 0, and subjected to a monochromatic parallel beam of light incident along the x axis (Fig. 25). The radiation energy incident on the semiconductor is equal to the product of the quantity $h\nu = hc/\lambda$ and the number of photons $N_{h\nu}$ falling on an area of 1 cm^2 of the semiconductor surface in 1 sec. The absorption coefficient will be denoted by α_λ and the reflectivity of the semiconductor surface by R_ν. We shall assume that a p-n junction is located at a distance x_s from, and is parallel to, the surface.

We shall denote the carrier diffusion coefficients by D_n and D_p and the pair diffusion coefficients (ambipolar diffusion coefficients) in the n- and p-type regions of the crystal by D_n' and D_p', respectively. The surface recombination velocity at x = 0 will be denoted by s. The diffusion lengths, representing volume re-

---

*The terms "collection coefficient" and "partition coefficient" are also used (cf., for example, [12]).

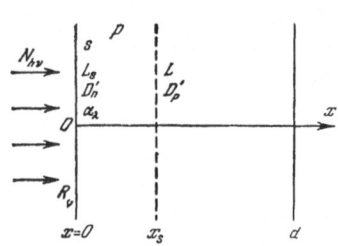

Fig. 25. Notation used in dealing with a photocurrent in a crystal with a p-n junction and with the quantum efficiency.

combination, will be denoted by L (in the n-type region) and L_s (in the p-type region). The carrier lifetime is $\tau = L^2/D'$. The following additional assumptions will be made:

a. the thickness of the space-charge region of the junction is small compared with x_s, L_s, and L;

b. an electric field exists only in the region of the junction;

c. the crystal thickness is $d \gg 1/\alpha_\lambda$, i.e., practically the whole photon flux $(1 - R_\nu)N_{h\nu}$ is absorbed in the semiconductor;

d. the hole and electron densities are sufficiently low so that we can use the Boltzman expression for the distribution function.

If the p- and n-type regions of the crystal are connected externally, the current in the external circuit is equal to the sum of the hole current J_p flowing from the n-type to the p-type region because of hole diffusion to the junction, and of the electron current J_n from the p-type to the n-type region. Denoting the current densities by j_p and j_n, respectively, the generation function by g, and the velocity of recombination of nonequilibrium carriers by u, we can write down the continuity equation for the steady-state conditions:

$$\left.\begin{aligned} -\frac{1}{q}\operatorname{div} j_n &= g - u, \\ +\frac{1}{q}\operatorname{div} j_p &= g - u. \end{aligned}\right\} \tag{2.15}$$

In the absence of a field

$$\left.\begin{aligned} j_n &= qD_n \operatorname{grad} n, \\ j_p &= -qD_p \operatorname{grad} p. \end{aligned}\right\} \tag{2.16}$$

In the one-dimensional case

$$\left.\begin{aligned} -D_n \frac{d^2 n}{dx^2} &= g - u, \\ -D_p \frac{d^2 p}{dx^2} &= g - u. \end{aligned}\right\} \tag{2.17}$$

In order to introduce the ambipolar diffusion coefficients, we shall multiply the first of the two equations in (2.17) by σ_{p0}, and the second equation by σ_{n0}. Adding, and dividing by $\sigma_{p0} + \sigma_{n0}$, we find, for the p-type region (on the assumption that the excitation level is low)

$$
\left.
\begin{aligned}
-D'_p \frac{d^2 \Delta n}{dx^2} &= g - u, \\
D'_p &= \frac{n_0 + p_0}{\dfrac{n_0}{D_p} + \dfrac{p_0}{D_n}}.
\end{aligned}
\right\}
\tag{2.18}
$$

A similar equation applies in the n-type region. Replacing u with $\Delta n/\tau$ in Eq. (2.18), we can write

$$
D'_p \frac{d^2 \Delta n}{dx^2} - \frac{\Delta n}{\tau} + g(x) = 0.
\tag{2.19}
$$

The boundary conditions for the p-type region are:

$$
n \big|_{x=x_s} = n_p e^{\frac{qV}{kT}},
\tag{2.20}
$$

$$
j_n \big|_{x=0} = qs(n - n_p) \big|_{x=0}.
\tag{2.21}
$$

In Eq. (2.21), the quantity s has the dimensions of cm/sec. The solution of Eq. (2.19) leads to the following expression for the quantum efficiency when the photoionization quantum yield is $\eta = 1$:

$$
\beta = \beta_1 + \beta_2 = \frac{\dfrac{1 - e^{\left(\frac{1}{L_s} - \alpha_\lambda\right) x_s}}{1 - \dfrac{1}{\alpha_\lambda L_s}} - \dfrac{s' - 1}{s' + 1} \dfrac{1 - e^{\left(-\frac{1}{L_s} - \alpha_\lambda\right) x_s}}{1 - \dfrac{1}{\alpha_\lambda L_s}}}{e^{\frac{x_s}{L_s}} - \dfrac{s' - 1}{s' + 1} e^{-\frac{x_s}{L_s}}} + \frac{e^{-\alpha_\lambda x_s}}{1 + \dfrac{1}{\alpha_\lambda L}},
\tag{2.22}
$$

where $s' = sL_s/D'_p$.

The first term, β_1, in Eq. (2.22) represents the quantum efficiency for carriers in the surface layer through which light passes into the semiconductor. It follows from Eq. (2.22) that β_1 depends on x_s which is the thickness of this layer (the depth of the p-n junction), on the surface recombination velocity, on the carrier diffusion length L_s, on the carrier diffusion coefficient D_p', and on the absorption coefficient α_λ. The second term, β_2, gives the quantum efficiency for the remainder of the crystal.

The complex expression in Eq. (2.22) must be used in those cases when photoionization occurs on both sides of a p-n junction, for example, in the analysis of the spectral characteristics of semiconductor "solar batteries" [14, 12].

In those cases when the spectral dependence of the quantum yield of germanium and silicon crystals could be investigated, the greatest interest lay in the value of the quantum yield within the fundamental absorption band. The following condition then applies:

$$\frac{1}{\alpha_\lambda} \ll x_s, \quad \frac{1}{\alpha_\lambda} \ll L_s, \tag{2.23}$$

and, consequently, the cumbersome expression for the quantum efficiency becomes

$$\beta = \beta_1' = \frac{2}{\left(1 + \frac{sL_s}{D'}\right)e^{\frac{x_s}{L_s}} + \left(1 - \frac{sL_s}{D'}\right)e^{-\frac{x_s}{L_s}}}. \tag{2.24}$$

In this case, the value of β is independent of the absorption coefficient.

It should be stressed that in stating the problem we have assumed that the semiconductor is homogeneous right up to its geometrical surface and that recombination near the surface is described by introducing a quantity s with the dimensions of cm/sec. We ignored the possibility of the appearance of an inversion layer at the surface or of a "dead" layer in which there is no pair generation.

Thus, illuminating a crystal with a p-n junction and measuring the external-circuit current with a low-resistance instrument, we can determine the photoionization quantum yield:

$$\eta = \frac{J_{pn}}{(1 - R)\, qN_{h\nu}\beta}. \tag{2.25}$$

The above expression indicates that apart from the value of the current J_{pn}, it is necessary to know the reflectivity R and the number of photons $N_{h\nu}$ incident on the semiconductor surface.

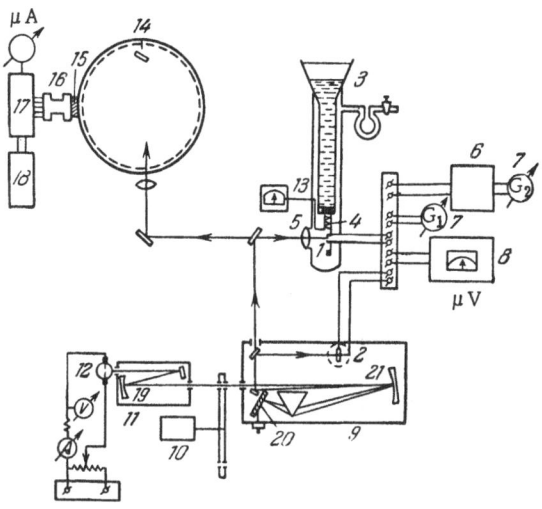

Fig. 26. Apparatus for the experimental investigation of photoionization in semiconductor crystals with p-n junctions [6]. 1) Test sample; 2) thermopile; 3) cryostat; 4) electrical furnace; 5) quartz lens; 6) potentiometer; 7) G_1 and G_2 are low-resistance galvanometers; 8) narrow-band amplifier; 9) mirror monochromator; 10) light chopper; 11) focusing system; 12) light source; 13) thermocouple and indicator; 14) holder with MgO surface or sample; 15) plate with phosphor; 16-18) photomultiplier with stabilized power supply; 19-21) aluminized spherical mirrors.

C. Results of Studies of the Spectral Dependence of the Photoionization Quantum Yield

In 1947, S. I. Vavilov indicated that at sufficiently high photon energies the quantum yield of photoluminescence might exceed unity [15]. The experiments of Butaev and Fabrikant confirmed this prediction [16].

According to the band theory of semiconductors, the initial kinetic energy of a photoelectron and a hole, both generated by the absorption of a photon, increases if the photon energy becomes greater than the forbidden bandwidth E_g. One would expect that an increase of the photoionization quantum yield due to excess photon energy would also affect the nonequilibrium carrier density. This was detected in germanium crystals by the

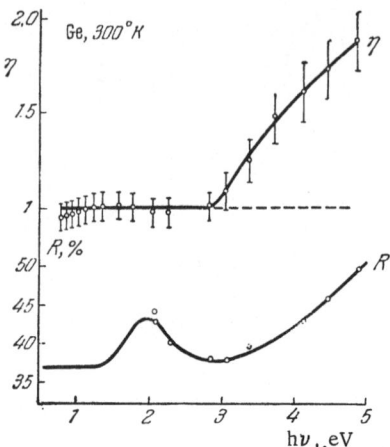

Fig. 27. Spectral dependence of the quantum
yield η in germanium excited with wavelengths
in the fundamental absorption band. R is the
reflectivity of the test-crystal surface.

Czech worker, S. Koc [17]. Using germanium and silicon crystals with p-n junctions and simple experimental conditions allowing the use of the above formulas, V. S. Vavilov and K. I. Britsyn
[18, 19] investigated the spectral dependence of the photoionization quantum yield for photon energies up to 4.9 eV (λ = 0.254 μ).
The basic circuit of the apparatus used for the measurements is
shown in Fig. 26. The intensity of radiation incident on the surface of a crystal with a p-n junction was determined with a cal-
ibrated thermopile 2. The method was described in greater detail
in [18, 19].

The results obtained for germanium and silicon crystals are
given below.

Germanium. Measurements were carried out on single-crystal n-type plates of resistivity ρ from 10 to 20 Ω·cm and having
an initial diffusion length L ≈ 1.5 mm. The crystals were 0.3-
0.6 mm thick. A p-n junction was produced by fusing in indium
on the side opposite to the illuminated surface. A second, nonrectifying contact was established by the simultaneous fusion of
tin. Tests were carried out at room temperature. A typical dependence of the quantum yield η on the photon energy hν and the

reflectivity spectrum of the crystal surface used in the calculation of η are shown in Fig. 27. The former curve has two clearly separate regions. In one of them, η is constant and equal to unity; in the second, the quantum yield increases with the photon energy. In the former region, the quoted data agree with the result of Goucher [9].

Silicon. Experimental data on the spectral dependence of the photoionization quantum yield of silicon single crystals with p-n junctions were published in 1958 [20] (cf. also [18]); p-type single crystals were used in this work. The junctions were produced by the thermal diffusion of phosphorus from the gaseous phase [14], currently widely used in the manufacture of Soviet silicon "solar batteries."

The dependences of the product $\eta\beta$ on the photon energy hν plotted in Figs. 28 and 29 are based on the data on the incident radiation flux, reflectivity, and photocurrent for all the inves- tigated silicon crystals. Depending on the state of the crystal surface and the depth of the junction, these curves can be divided into two types. Curves of one type (Fig. 28), for thin n-type lay- ers and "clean" (etched) surfaces, exhibit a plateau in the region of low values of hν (approximately up to 3.2 eV); in the region of this plateau, the value of $\eta\beta$ is constant but beyond it the product $\eta\beta$ increases with increase in hν. Curves of this type can be interpreted directly by means of Eqs. (2.24) and (2.25), because in the case of sufficiently strong absorption, i.e., when $1/\alpha_\lambda \ll x_S$, the value of β is independent of the absorption coefficient α_λ, as indicated by Eq. (2.24). From the results of a direct determina- tion of the value of α_λ in the fundamental absorption region of silicon [21], and from the value of α_λ estimated from the reflec- tivity [19], it follows that at photon energies greater than 2.5 eV $1/\alpha_\lambda < 10^{-5}$ cm, and for hν = 4.9 eV we have $1/\alpha_\lambda < 10^{-6}$ cm. Thus, even in the visible region, and particularly in the ultra- violet region, the value of β is independent of hν for d $\approx 10^{-4}$ cm.

However, as just pointed out, in real crystals β can be in- dependent of hν only in the absence of a "dead" layer near the surface. In this "dead" layer, photons may be absorbed without carrier generation, which has not been allowed for in the deriva- tion of expressions such as Eq. (2.24).

In our opinion, $\eta\beta = f(h\nu)$ curves of the second type (Fig. 29), which exhibit a "dip" on the right of the plateau or a strongly non- linear rise of $\eta\beta$ with hν, indicate the presence of a "dead" layer

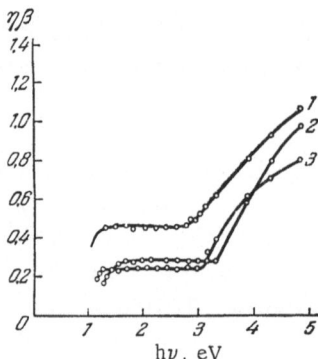

Fig. 28. Dependence of the product $\eta\beta$ on the photon energy $h\nu$ for silicon. 1) 100°K; 2) 300°K; 3) 400°K (type 1).

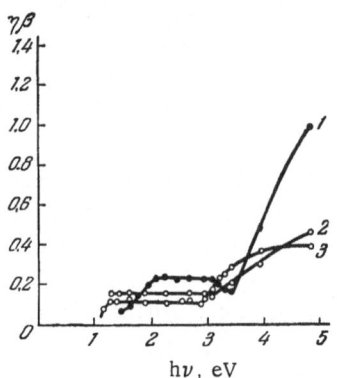

Fig. 29. Dependence of the product $\eta\beta$ on the photon energy $h\nu$ for silicon. 1) 100°K; 2) 300°K; 3) 400°K (type 2).

at the surfaces of the test samples. This is confirmed by the spectral dependence of the characteristics of silicon solar batteries [14] and by the results of experimental studies of ionization in silicon when it is bombarded with fast electrons [19, 22].

Analysis of the spectral characteristics of the "second type" coupled with the independent determination of the surface recombination velocity may give information on the "dead" layer. Such information is of intrinsic interest in connection with the problems of increasing the efficiency of photocells and the efficiency of recording and counting short-range charged particles by means of silicon detectors.

From the point of view of investigating the primary photoionization process, the most important are the spectral characteristics of those silicon crystals for which the region of constant $\eta\beta$ is wide and changes without a dip to a region of increasing $\eta\beta$. It is understood that in this case the influence of "dead" layers is not eliminated. However, in practice, this influence is restricted to the region of $\eta\beta$ increasing with $h\nu$, where it reduces the product $\eta\beta$; therefore the presence of a "dead" layer simply reduces the value of the photocurrent.

Using the accepted ideas on the generation of pairs by the absorption of photons, we may assume that in the energy region $E_g < h\nu < 2E_g$, the quantum yield is equal to unity, which is con-

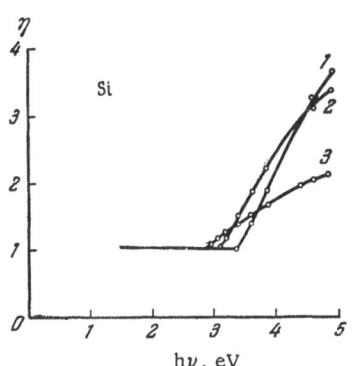

Fig. 30. Spectral dependence of the photoionization quantum yield η for silicon: 1) 100°K; 2) 300°K; 3) 400°K.

firmed by the presence of a plateau. Extrapolating the constant value of β to the region where the product $\eta\beta$ increases, we can plot $\eta = f(h\nu)$, i.e., the spectral dependence of the absolute value of η (Fig. 30).

Apart from germanium and silicon, the increase in the photoionization quantum yield has been observed in indium antimonide [11], cadmium telluride (CdTe) [23], lead sulfide, and other semiconductors. Figure 31 shows the spectral dependence of the quantum yield of InSb found by measuring the photoconductivity and photomagnetoelectric effect, the latter being used to determine independently the nonequilibrium pair lifetime.

D. Discussion and Interpretation of the Data on the Spectral Dependence of the Photoionization Quantum Yield

The increase of the quantum yield at relatively high photon energies can be explained most simply by the process of impact ionization, i.e., by the generation of secondary carrier pairs at the expense of the excess kinetic energy of a photoelectron of a photohole. Obviously, this secondary ionization is the consequence of the interaction of photoelectrons with valence electrons of the host atoms and not of the impurity atoms. This is indicated both by the absolute values of η, determined experimentally, and by the fact that the rise in η occurs at room temperature when the atoms of the usual donor or acceptor impurities in Ge and Si are almost completely ionized. Moreover, the rise in η with $h\nu$ is observed only at values of $h\nu$ considerably greater than the forbidden bandwidth. According to the theory of McKay [24] and Wolfe [25] (which they used to interpret data on impact ionization in Ge and Si), in a strong electric field when the kinetic energy of carriers in a band is greater than a certain threshold value E_i, the probability of impact ionization should be greater than the probability

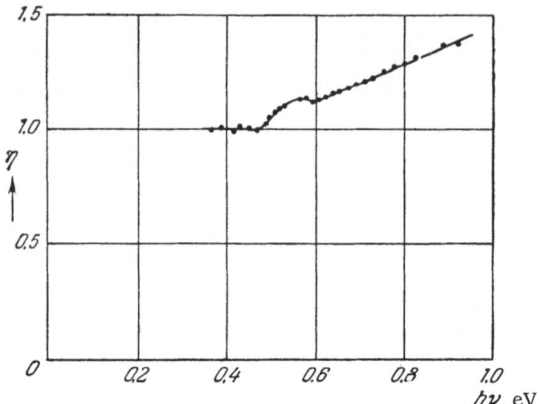

Fig. 31. Spectral dependence of the photoionization
quantum yield of indium antimonide according to Tauc
[11].

of energy loss by any other method. According to the data of
McKay, this threshold energy is $E_i \approx 2.3$ eV for silicon. Thus,
in accordance with this interpretation, the quantum yield should
rise steeply to a value of 2, beginning from $h\nu \approx E_g + E_i$, which
disagrees with the experimental data. On the other hand, according to this theory, η cannot increase to a value greater than
2 without generating another pair which should be represented by
a further "kink" in the $\varphi = f(h\nu)$ curve. However, such a kink
has not been found for germanium or silicon.

In our opinion, bearing in mind the small difference between
the effective masses of electrons and holes in silicon, we should
assume that the excess kinetic energy is equally divided between
the two carriers composing the primary pair. Consequently, one
would expect a considerable rise in the quantum yield beginning
at about $h\nu \approx 3E_g$, i.e., approximately at 3.3 eV for silicon at
room temperature, as observed experimentally [26].

A similar treatment but in a more detailed form was recently
given by W. Shockley [7]. Having assumed that the energy obtained by the absorption of a photon is divided equally between two
primary carriers, and stating that the minimum energy needed for
impact ionization is E_g, Shockley considered quantitatively the
relationships between the probabilities of the competing processes
of secondary ionization and emission of a phonon of maximum energy E_R. Using the latest data on the phonon energies in silicon

and germanium, obtained from neutron-scattering data, Shockley plotted theoretical curves of the spectral dependence of the quantum yield, and found that they were in good agreement with experiment in the case of silicon and in satisfactory agreement in the case of germanium. The only parameter whose value was selected was the average number of phonons n generated by a photoelectron before the creation of a secondary pair. It was found that in the case of silicon at 300°K this number was n = 17.5. Before Shockley, the process of impact ionization by photoelectrons was considered by Antoncik [27], who gave a quantitative explanation of the experimental data for Ge and Si without allowing for the energy distribution between a photoelectron and a photohole.

In the region where the quantum yield is considerably greater than unity, its value is determined by the relationship between the probabilities of impact ionization and the interaction of "hot" electrons or holes with the crystal. Consequently, it was of interest to obtain data on the process of photoionization at very different temperatures. According to the results for silicon [19], the rise of $\eta\beta$ and η with hν is strongly temperature dependent (Fig. 30). From typical curves it is evident that at high temperatures (400°K) the quantum yield rise occurs at considerably lower photon energies, beginning with 2.95 eV, compared with 3.4 eV at 100°K. This displacement of the beginning of the quantum yield rise is partly due to the narrowing of the forbidden band with increase in temperature: at 100°K, E_g for silicon is 1.17 eV, and at 400°K, it is 1.04 eV.

Since the generation of the primary and secondary carrier pair requires each time an energy $E \geq E_g$, the narrowing of the forbidden band can explain the displacement of the quantum yield rise by an amount $2 \times (1.17 - 1.04)$ eV = 0.26 eV. This leaves unexplained a further displacement of about 0.2 eV, which is more than the experimental error and is not accounted for by current theories. We may assume that this displacement is due to an increase in the relative probability of "indirect" electron transitions involving phonon participation.

§ 11. Ionization by Absorption of High-Energy Photons (X Rays and Gamma Rays)

The passage of x rays or gamma rays through a crystal is accompanied by the formation of fast electrons due to the photoeffect

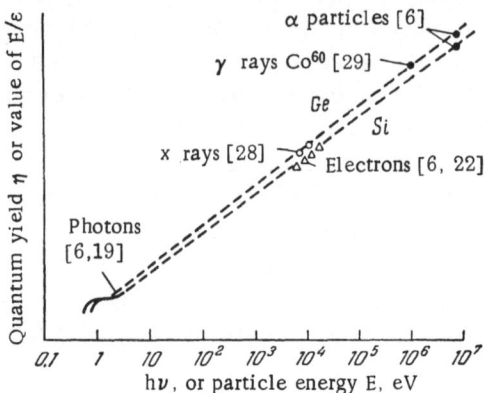

Fig. 32. Experimental data on the ionization of germanium and silicon by photons and charged particles. The
upper curve represents germanium, the lower one silicon.

or the Compton effect. We shall restrict ourselves to the range
of energies in which the formation of electron-positron pairs may
be neglected compared with the photoeffect and the Compton effect.

Under the usual experimental conditions, almost all the fast
electrons stay in the crystal and their energy is used up in the
formation of a large number of excess-carrier pairs and in the
excitation of crystal-lattice vibrations. *

The concept of the photoionization quantum yield includes in
this case a complex chain of intermediate processes. Current
ideas on this process allow us to assume that the average energy
ε used to form one pair of excess carriers   should be independent of the primary photon energy $h\nu$ or of the initial fast-electron
energy [6]. Photocurrents produced by monochromatic x rays
with $h\nu$ = 10 keV and 20 keV in single crystals of germanium with
p-n junctions have been investigated by the Czech physicists,
Drahokoupil, Malkouska, and Tauc [28]. M. V. Chukichev and
V. S. Vavilov investigated the ionization of Ge single crystals by
Co^{60} gamma rays [29]. Values of the average ionization energy
ε, determined by these two groups of workers, were identical, and
equal to 2.5 ± 0.3 eV. Thus, at high photon energies $\eta = h\nu/\varepsilon$
in the case of germanium (Fig. 32).

* The problem of ionization by fast electrons and other charged particles will be considered separately in Chapt. III.

Summarizing the available data on the photoionization quantum
yield near the fundamental absorption edge and at high photon energies, we can suggest that not only Ge and Si but also other semiconductors should have a region in which the quantum yield is constant and equal to unity, extending from the fundamental absorption
edge to photon energies of about $3E_g$, followed by a nearly linear
rise of the "quantum yield" η with the photon energy.

The cause of the rise in η is the impact ionization caused by
photoelectrons and photoholes which have considerable kinetic
energy at the moment when they are generated.

§ 12. Recombination and Capture of Electrons and Holes in Semiconductors

In the preceeding chapter, we discussed the processes of the
absorption of light by semiconductors. At the beginning of the
present chapter, we introduced the principal concepts necessary to
describe the photoconductivity and we discussed the primary photoionization process.

We showed that in the case of a given generation function the
steady-state value of the photoconductivity in a homogeneous semiconductor is governed by the nonequilibrium carrier lifetimes.
Theoretical and experimental investigations of recombination processes represent an important branch of the physics of semiconductors, because they are of interest outside the limits of "radiation physics, " which is taken to mean the initial interaction of radiation with crystals. We shall restrict ourselves to discussing
several typical problems in the theory of recombination, which
answer the following questions in photoconductivity (or, more correctly, radiation conductivity):

1. Which processes restrict the value of the lifetime and,
consequently, of the steady-state photoconductivity?

2. How does the volume recombination velocity, governed
by one type of impurity recombination center, depend on the equilibrium electrical conductivity?

3. How do the trapping centers (traps) affect the steady-state
value and kinetics of the photoconductivity?

4. What are the reasons for the nonlinear dependence of the
photoconductivity on the excitation intensity?

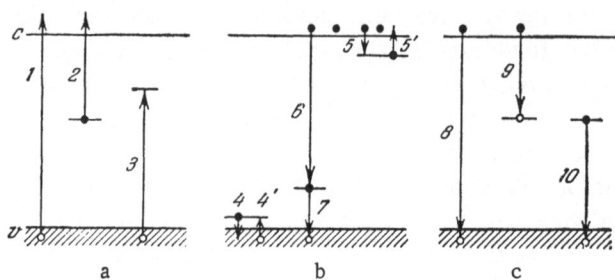

Fig. 33. Electron transitions in a semiconductor with local centers:
a) Photoionization; b) carrier capture; c) recombination.

5. What is the explanation for the "quenching" of photoconductivity, i.e., the decrease of photoconductivity excited by light of one wavelength under the action of additional illumination with light of another wavelength?

6. What is the explanation of the phenomenon of negative photoconductivity, i.e., of the reduction of the electrical conductivity as a result of the absorption of radiation energy?

The phenomena of carrier capture and recombination, associated with local centers and defects, can be conveniently studied using the system of electron transitions shown in Fig. 33. The transitions 1, 2, and 3 (Fig. 33a) represent the absorption of light in the fundamental band (1) and by localized impurities (2 and 3). In case 1, a free-carrier pair is generated; in case 2, a free electron and a bound (localized) hole; in case 3, a free hole and a bound electron. Electron transitions representing the generation of excitons or excited impurity centers, and intraband transitions are not shown, since they are not accompanied by the generation of free carriers.

The nonequilibrium electrons and holes, produced as a result of photoionization, exist until they are captured by impurity centers. Usually, the capture is more likely than the direct recombination or the formation of excitons. The centers (traps) capable of capturing free carriers are frequently divided into:

a. trapping centers – if the captured carrier has a higher probability of being thermally excited back to the free state compared with the probability of recombining at the center with a carrier of opposite sign, and

b. recombination centers – if the captured carrier is more likely to recombine with a carrier of opposite sign.

In general, a center with an energy level close to one of the bands usually acts as a trapping center, and a center close to the middle of the forbidden band acts as a recombination center. The difference between trapping and recombination centers lies in the ratio of the probabilities of thermal liberation and recombination.

The transitions 4 and 4', 5 and 5' in Fig. 33b represent the capture and thermal liberation of carriers; the transition 6 represents the capture of an electron by a recombination center; and the electron transition 7 represents the capture of a hole.

At one temperature or excitation level, a center may act as a trapping center, and at another temperature or excitation level, it may act as a recombination center [58].

The system shown in Fig. 33c represents the three main types of electron transition in the case of recombination. First, a free electron may recombine directly with a free hole (transition 8). Transitions of this type are usually radiative, i.e., they are accompanied by the liberation of energy in the form of a photon whose energy is then approximately equal to the width of the forbidden band. Such radiation is frequently called "edge" radiation. The probability of this direct recombination is usually very low and does not affect the average values of the nonequilibrium carrier lifetimes.

More likely, recombination processes are represented by the transitions 9 and 10, i.e., by the capture of an electron by a center near which there is a bound hole, the capture of a hole by a center near which an electron is localized. These two transitions may also be radiative. Thus, the return of the electron system of a crystal from an excited state to the equilibrium state may be accompanied by luminescence, which is also known as the "recombination radiation" in semiconductors (cf. Chapt. IV). The probability, i.e., the "velocity" of recombination is governed mainly by the method of the transformation of the excess energy of the excited "nonequilibrium" carriers. The energy of nonequilibrium carriers can be transformed or dissipated in various ways:

a. by the emission of light (photons),

b. by the transfer of the excitation energy to the crystal lattice, i.e., by the emission of phonons,

c. by the transfer of the excess energy of two recombining carriers to a third carrier, i.e., a process which is the converse of impact ionization. This process is known as "impact recombination" or the Auger effect.

Naturally, two or more different energy transfer processes may occur simultaneously: for example, we may have the simultaneous emission of a photon and the excitation of phonons.

When several phonons are generated, they may appear either simultaneously or consecutively (as a cascade) because a carrier may approach a local center and then move near it either emitting or absorbing phonons [30] until it is captured or has moved away from the center.

The lifetimes corresponding to the capture of carriers by hydrogen-like impurity centers in germanium crystals were calculated by Sclar and Burstein [31], and Gummel and Lax [32], using the formula

$$\tau = \frac{1}{C_T N_r},$$ (2.26)

where N_r is the concentration of the centers. They assumed that the total probability of carrier capture C_T is given by

$$C_T = C_{rad} + C_{phon} + nC_{impact}$$

where n is the free carrier density. According to these authors [31, 32], at low temperatures the probability of capture accompanied by phonon emission is much lower than the probability of capture involving energy transfer to the lattice (see also Chapt. IV). The probability of impact recombination should be highest at high temperatures. However, a comparison of the theoretical calculations with the experimental data on the lifetimes in Ge and Si showed that the theoretical values of C_{impact} are several orders too high.

A. Recombination Capture of Carriers by Local Centers

In the majority of cases, the nonequilibrium carrier lifetime in semiconductors is governed not by direct recombination but by the presence of defects, especially local centers. In contrast to direct recombination, whose velocity is determined by the carrier density and one constant (the probability), the process of recombination at local centers is also influenced the carrier population of the centers. The degree of population is related to the free-carrier density through the probability (the effective cross section) of electron capture by a vacant center and through the

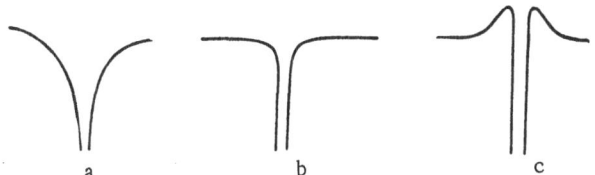

Fig. 34. Schematic representation of the potential distribution near capture centers: a) Coulomb attraction; b) neutral centers; c) Coulomb repulsion.

effective cross section for the capture of a hole by a center at which an electron is localized.

If unit volume contains N recombination centers, then the lifetime of a free carrier in the case of capture is given by

$$\tau = \frac{1}{vSN},\qquad(2.27)$$

where $v = \sqrt{2kTm^{*-1}}$ is the velocity of thermal motion of the carrier and S is the effective capture cross section.

The value of S is determined by the nature of the variation of the potential near the center (cf. Fig. 34). Experiments show that centers whose electric charge attracts free carriers of opposite sign have large effective cross sections. It is usually assumed that a center captures a free carrier if the latter approaches so closely that the binding energy becomes equal to the average thermal excitation energy kT. Thus, for a center with a single charge

$$\frac{q}{\varepsilon r} = kT,\qquad(2.28)$$

i.e., at room temperature $S = \pi r^2 \approx (10^{-10}/\varepsilon^2)\,cm^2$; in the case of Ge, Si, and several other substances whose ε is of the order of 10, $S \approx 10^{-12}\,cm^2$. For a neutral center, one would expect values of S close to $10^{-15}\,cm^2$, which is of the order of the "atomic cross section," in agreement with the experimental data.

On the other hand, a center whose charge is of the same sign as that of a free carrier is surrounded by a potential barrier which causes electrostatic repulsion. Such centers should have very low capture cross sections.

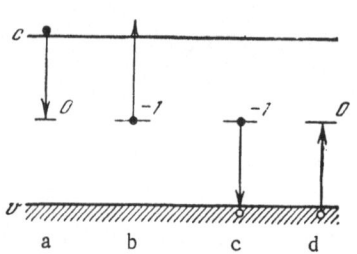

Fig. 35. Four types of electron transition to be allowed for in calculating the recombination velocity.

B. Statistics of Recombination Processes for Centers with One Level

In many cases, a crystal has one type of dominant recombination center and the nonequilibrium lifetime is governed by the processes of electron capture and subsequent hole capture by the local levels of the dominant centers. The statistics of recombination capture by centers with one level in the forbidden band was considered by R. H. Hall [33], and by W. Shockley and W. Read [34]. Their analysis gave a dependence of the lifetime on the Fermi level position if the recombination level position and the electron and hole capture cross sections were known; in some cases, they were able to determine the recombination center parameters from the temperature dependence of the lifetime.

The arrows in Fig. 35 represent four electron transitions which must be allowed for in the calculation of the recombination velocity. In case (a) a neutral center captures an electron; (b) represents the transition of an electron to the conduction band; (c) represents the capture of a hole by a negatively charged center; (d) is the transition of a hole from a neutral center to the valence band.

The absolute value of the rate of capture of free electrons u_{cn} is given by the formula:

$$u_{cn} = n\,(N_r - n_r)\,vS_n - n_r N_c vS_n \exp\left(-\frac{E_r}{kT}\right), \qquad (2.29)$$

where N_r is the concentration of recombination centers; n_r is the concentration of recombination centers filled with electrons; v is the average velocity of the thermal motion of carriers; S_n is the cross section for the capture of free electrons by recombination centers. The ionization energy of a recombination center is denoted by E_r; N_c represents the effective density of states in the conduction band.

The degree of electron population at equilibrium is determined by the Fermi level position; if we denote the absolute value of the energy gap between the bottom of the conduction band and the Fermi level by E_F then

$$n_r = N_r \frac{1}{1 + m \exp\left(\dfrac{E_F - E_r}{kT}\right)}. \qquad (2.30)$$

Denoting the free-electron density when $E_F = E_r$ by n', and the ratio of the degree of degeneracy of the states (which, for the sake of simplicity, we shall take to be unity) by m, then

$$n' = N_c \exp\left(-\frac{E_r}{kT}\right) \qquad (2.31)$$

and

$$u_{cn} = nN_r (1 - f_r) v S_n - N_r f_r n' v S_n, \qquad (2.32)$$

where

$$f_r = \frac{1}{1 + \exp\left(\dfrac{E_F - E_r}{kT}\right)}. $$

Similarly, the rate of capture of holes u_{cp} can be written as follows:

$$u_{cp} = pN_r f_r v S_p - N_r (1 - f_r) p' v S_p, \qquad (2.33)$$

where p is the density of holes, p' = $N_v \exp[-(E_g - E_r)/kT]$, E_g is the forbidden bandwidth, and N_v is the effective density of states in the valence band.

Under steady-state conditions when the rate of generation of nonequilibrium electron-hole pairs is u, we should have the equality

$$u = u_{cn} = u_{cp}. \qquad (2.34)$$

Equating the expressions for u_{cn} and u_{cp}, we can express the Fermi distribution functions f_r in the form

$$f_r = \frac{S_n n + S_p p'}{S_n (n + n') + S_p (p + p')}. \qquad (2.35)$$

Introducing, for brevity, the capture probabilities

$$C_n = N_r v S_n \quad \text{and} \quad C_p = N_r v S_p \qquad (2.36)$$

and substituting the value of f_r into Eq. (2.32) or Eq. (2.33), we obtain the expression

$$u = \frac{C_n C_p (pn - p'n')}{C_n (n + n') + C_p (p + p')}. \qquad (2.37)$$

The product p'n' can be expressed as follows

$$p'n' = N_c N_v \exp\left(-\frac{E_g}{kT}\right) = n_i^2, \qquad (2.38)$$

where $n_i = p_i$ is the equilibrium density of free electrons or holes in a material with intrinsic conductivity. Moreover, to make the case definite, we shall assume that

$$E_r < E_i = \frac{E_g}{2} + \frac{1}{2} kT \ln \frac{N_v}{N_c}, \qquad (2.39)$$

where E_i is the energy corresponding to the Fermi level position in a material with intrinsic conductivity. It follows that $n' > n_i > p'$. If $E_r > E_i$, then the quantities corresponding to electrons and holes should be interchanged.

We shall consider a typical case: let us assume that the excess carrier density δn is small compared with the density of majority carriers at thermal equilibrium; the electron lifetime is equal to the hole lifetime, i.e., we can neglect the change in the bound charge when the centers become filled.

The total carrier densities n and p are

$$n = n_0 + \delta n,$$

$$p = p_0 + \delta p \qquad (2.40)$$

(n_0 and p_0 are the equilibrium densities). The lifetime can be found directly from the expression

$$\tau \equiv \frac{\delta n}{u}. \qquad (2.41)$$

Substituting the expressions for n and p into Eq. (2.37), we obtain τ in the form:

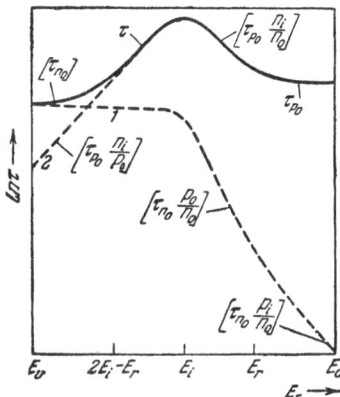

Fig. 36. Dependence of the carrier τ on the Fermi level position for a fixed concentration of recombination centers.

$$\tau = \tau_{p0} \frac{n_0 + n'}{n_0 + p_0} + \tau_{n0} \frac{p_0 + p'}{n_0 + p_0}. \quad (2.42)$$

The quantity $\tau_{p0} \equiv 1/C_p$ represents the value of the hole lifetime in a material which is strongly n-type, and $\tau_{n0} \equiv 1/C_n$ is the electron lifetime in a material which is strongly p-type. Figure 36 gives the dependence of the nonequilibrium carrier lifetime on the Fermi level position in the case of weak excitation; this dependence is given by Eq. (2.42) for a semiconductor with a fixed concentration of recombination centers having one capture level for different positions of the Fermi level; thus the portions of the curve $\tau = f(E_F)$ corresponding to E_F close to E_V represent crystals strongly doped with acceptor centers, and the portions where E_F is close to E_C represent crystals with a high concentration of donors. Curve 1 represents the first term in Eq. (2.42), curve 2 the second term, and the continuous curve gives the resultant value of τ.

It is evident from Fig. 36 that a material which is strongly p-type is that material for which the Fermi level lies between E_V and $2E_i - E_r$. In this case, all the recombination centers are vacant and the lifetime is governed only by the probability of electron capture.

In the case of a material whose p-type nature is less pronounced and the Fermi level lies between $2E_i - E_r$ and E_i, τ increases with the approach of the Fermi level to E_i as follows

$$\tau = \tau_{p0} \exp \frac{E_g - E_r - E_F}{kT}. \quad (2.43)$$

Many centers are unfilled but there is now a high probability that the captured electron will be ejected back to the conduction band to recombine eventually with a free hole.

In a material which is weakly n-type, i.e., when the Fermi level lies between E_i and E_r, τ decreases with the approach of the Fermi level to E_r:

$$\tau = \tau_{p0} \exp\left(\frac{E_P - E_r}{kT}\right). \tag{2.44}$$

In this case, many centers are still not filled with electrons and the recombination is governed only by the ability of the electron-filled centers to capture holes.

Finally, in a material which is strongly n-type, the Fermi level lies between E_r and E_C and $\tau \to \tau_{p0}$; practically all the centers are occupied by electrons and the recombination velocity is governed by the probability of the capture of free holes.

C. A Typical Recombination Process at Local Centers:
Chemical Impurities in Germanium Crystals

J. Burton et al. [35] and also S. G. Kalashnikov et al. [36], showed that under certain conditions the impurities which form centers with deep levels in germanium crystals act as centers with one capture level; then the recombination process can be analyzed quantitatively using the Hall-Shockley-Read statistics.* Investigations of the role of copper and nickel atoms as recombination centers in Ge provided convincing proof of the strong influence of small concentrations of known impurity on the lifetime of nonequilibrium carriers and, consequently, on the photosensitivity of a semiconductor. The results of Burton et al. [35] are shown in Fig. 37. The upper curve, drawn through the experimental points (triangles), represents the lifetime in control single crystals of "pure" p- and n-type germanium. Maximum values of the lifetime, of the order of several milliseconds, correspond to the purest single crystals and are limited by the capture of carriers by dislocations, defects, and accidental impurities. The lower curve represents a large series of germanium single crystals doped with copper. Each of the circles through which the lower curve is drawn represents the nonequilibrium carrier lifetime measured in a germanium crystal grown from a melt con-

---

* Usually, an impurity has several deep energy levels corresponding to different charged states. W. Shockley and T. Sah [37], and S. G. Kalashnikov [38] generalized the theory of recombination to multicharged centers. Experiments confirmed the validity of their conclusions for nickel in germanium [39] and in certain other cases. However, if the ionization energies of the levels are sufficiently different, one can frequently use the simple one-level model to analyze the experimental data.

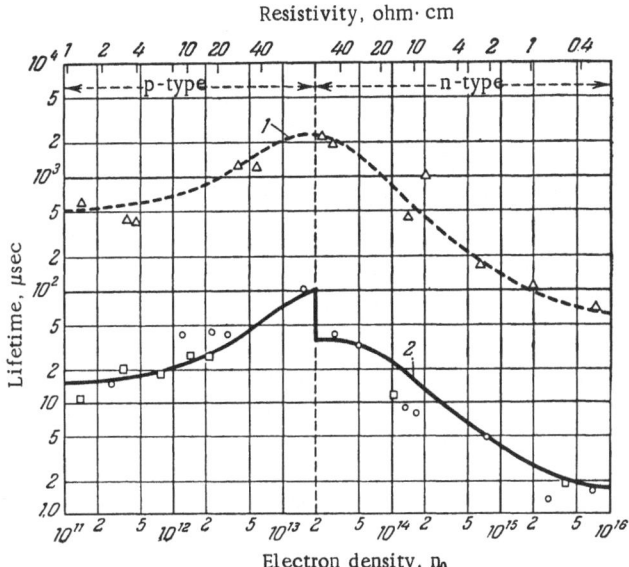

Fig. 37. Lifetimes of nonequilibrium carriers in copper-doped germanium crystals. Curve 1: pure material; curve 2: material
doped with copper.

taining copper impurity; the concentration of copper in these crystals was 3.6×10^{14} cm^{-3}. The squares represent the lifetimes in
crystals where the same concentration of copper was obtained by
thermal diffusion from the crystal surface. Within the limits of
the experimental error, the lifetimes were independent of the
method of doping with copper　The continuous curve drawn through
the circles and squares corresponds to the expression (2.42), which
follows from the analysis of the recombination statistics for one
level with $N_T = 3.6 \times 10^{14}$ cm^{-3}, $n' = 2.5 \times 10^{12}$ cm^{-3}, $S_n = 0.1$
$\times 10^{-16}$ cm^2, and $S_p = 1 \times 10^{-16}$ cm^2. The concentration of copper atoms was determined by two independent methods: from the
Hall effect and by a tracer method. It was assumed that the concentration of recombination centers was equal to the total concentration of copper atoms in the crystals.

It follows from Eqs. (2.42)-(2.44) that, without additional
information, an analysis of the dependence of the lifetime on the
Fermi level position cannot be used to find in which half of the
forbidden band the recombination level lies. In the case of copper
in germanium, the Hall effect data made it possible to identify the

recombination capture level with the deep acceptor level of copper at 0.3 eV from the valence band.

Apart from chemical impurities, which form capture centers with deep levels, point defects (interstitial atoms and vacancies) and dislocations can act as recombination centers. Up to the present, the influence of dislocations and defects on the velocity of recombination of electrons and holes has been investigated in greatest detail for germanium and silicon.

Okada [40] and Kulin [41] found that the lifetime in germanium single crystals depends considerably on the dislocation density: Ge crystals with fewer dislocations had longer lifetimes [40] and the quantity $1/\tau$ was proportional to the disolocation density. The more careful investigations of Milevskii [63] have shown that in Ge and Si crystals the recombination impurities such as Cu and Ni precipitate heavily on dislocations and thus, at a given residual impurity concentration, the maximum lifetimes are not exhibited by dislocation-free crystals but by crystals in which the maximum mutual neutralization of impurity centers and dislocations has taken place. In crystals free of chemical recombination centers, the carrier lifetime increases with improvement in the degree of perfection of the structure and, in the limit, may be determined simply by direct (radiative) recombination and unavoidable surface processes which should be considered as irremovable "defects."

D. Recombination of Nonequilibrium Carriers in Silicon and in Indium Antimonide

As for germanium, the recombination velocity in silicon single crystals is restricted by the presence of centers with deep levels. Such centers may be gold, copper, manganese, or iron atoms, each of them having several levels in the forbidden band [42]. It is assumed that in the case of gold in p-type silicon the recombination proceeds mainly through a donor level of Au with an electron capture cross section close to 4×10^{-15} cm^2 at 300°K, while in n-type silicon an acceptor level of Au with a hole capture cross section of 2×10^{-15} cm^2 is active. Heat treatment, neutron, fast-electron,or gamma-ray bombardment of silicon, all of which disturb the structure, are accompanied by the formation of recombination centers.

The lifetimes of nonequilibrium carriers in InSb were determined by investigating the photoconductivity and the photomagnetoelectric effect [43] and, more simply, by measuring the excess
conductivity decay after the sudden termination of excitation with
fast electrons accelerated by an electrostatic generator [44]. Indium antimonide is an interesting example of a semiconductor with
a narrow forbidden band (0.18 eV). According to the data of Wertheim [44], the lifetime in indium antimonide at low temperatures
(130-200°K) is governed by recombination centers; at higher temperatures, when intrinsic conduction predominates, direct (radiative) recombination plays the major role due to the narrow width
of the forbidden band (cf. Chapt. IV).

E. Processes Connected with the Capture (Trapping) of Carriers

The capture of free carriers by local centers (traps), not
accompanied by subsequent recombination, frequently determines
the kinetics and the steady-state value of the photocurrent. The
equation describing the time dependence of the density n_t of electrons captured by the local centers (traps of one type only) can
be written in the following way:

$$\frac{\partial n_t}{\partial t} = S_n v n \left(N_t - n_t\right) - n_t N_c v S_n \exp - \frac{E_t}{kT}, \qquad (2.45)$$

where N_t is the concentration of traps having energy levels at a
distance E_t from the conduction band. The expression (2.45) is
identical with Eq. (2.20).

In order to illustrate graphically the action of traps, we shall
introduce the following simplifying assumptions. We shall assume
that the steady-state value of the free-electron density is fixed; in
this case Eq. (2.45) becomes linear. Moreover, we shall assume
that the probability of the recombination of minority carriers is
much lower than the probability of capture by traps. The more
general case has been discussed by Haynes and Hornbeck [45].

The second of the above assumptions restricts our case to
the single capture of carriers by traps. When these conditions
are satisfied the solution of Eq. (2.45) leads to the relationships

$$S_n v = \frac{1}{n_e} \left(\frac{1}{\tau_0} - \frac{1}{\tau_g}\right), \qquad (2.46)$$

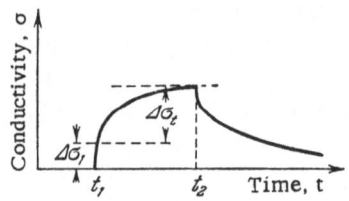

$$\frac{n_t^\infty N_c}{n_e\left(N_t - n_t^\infty\right)} = \exp - \frac{E_t}{kT}, \quad (2.47)$$

$$\tau_g = \frac{1}{N_c S_n v} \exp - \frac{E_t}{kT}. \quad (2.48)$$

Fig. 38. Photoconductivity rise and decay curves for a crystal with traps: t_1 is the moment when illumination begins, t_2 is the moment when it stops; $\Delta\sigma_1$ is the additional conductivity due to the excess carrier density n_e; $\Delta\sigma_1 + \Delta\sigma_t$ is the steady-state value of the photoconductivity.

The quantity τ_g is the average time after which the captured electron is liberated again; n_t^∞ is the steady-state value of the density of captured electrons, corresponding to the bottom of the conduction band, N_c is the effective density of states in that band and τ_0 is the time constant characterizing the capture of electrons by traps.

A typical photoconductivity rise and decay curve for a crystal with traps is shown schematically in Fig. 38. Immediately after the start of the optical excitation, the density of free electrons rises by the quantity n_e in a time of the order of τ, determined by recombination. The conductivity of the sample increases due to the excess density of electrons and holes by

$$\Delta\sigma_1 = q\mu_p n_e (1 + b), \quad (2.49)$$

where b is the ratio of electron and hole mobilities. The excess electrons increase the rate of capture by the traps, which are then gradually filled. The rate of filling of the traps can be varied by altering the intensity of the exciting light.

The electrons captured by traps do not take part in charge transport. However, the excess holes, needed to neutralize the captured electron charge, increase the electrical conductivity; this effect is represented in Fig. 38 by the quantity $\Delta\sigma_t$. The value of n_t can be calculated from the relationship

$$\Delta\sigma_t = q\mu_p n_t. \quad (2.50)$$

After the excitation is stopped, the density of captured electrons and the excess conductivity (due to the compensating holes) both decay.

The photoconductivity rise and decay curves are frequently used to determine the lifetime of nonequilibrium carriers. In the presence of traps, the time constants representing the filling or emptying of traps are sometimes erroneously identified with the lifetimes of electron-hole pairs. Usually, such errors can be eliminated by illuminating the test sample with light of sufficiently high constant intensity on which are superimposed exciting light pulses used to determine the photocurrent rise or decay time constants. The constant excitation allows us to maintain a sufficiently high degree of trap population and under such conditions it is frequently possible to separate out the time constants representing recombination. However, when the lifetime depends on the minority carrier density, * the rise and decay curves become nonexponential and the concept of the lifetime itself can be used only approximately.

It is evident from Eq. (2.46) that in the case of single capture the variation with time of the trapping level population is represented by a single time constant. However, it has been shown, for example for single crystals of silicon, that multiple capture is possible so that the probability of electron capture from the conduction band exceeds the probability of recombination.

In the latter case, the excess carrier density decreases fastest at the beginning when a considerable fraction of the traps is filled; later the photoconductivity decay approaches the exponential law. The rate of loss of excess carriers is governed by the relationship between the probabilities of capture by traps and by recombination centers [45].

Thus, although the kinetics of the photoconductivity of a semiconductor with traps does depend primarily on the trap parameters, the characteristic times cannot be assumed to be simple functions of temperature and of the trap parameters.

The processes of capture by traps (the "trapping" of carriers) are observed as a rule in all semiconductors exhibiting high photosensitivity. † Good examples of such semiconductors are cadmium sulfide and selenium. However, the nature of traps and their characteristics (the level depth and capture cross section) have

---

* See, for example, the review of S. M. Ryvkin [1].

† The semiempirical rule of proportionality between the photosensitivity and the slowness of the response is well known in the manufacture of photoresistors.

not yet been sufficiently studied. Relatively detailed data on the
energy levels acting as traps are available for germanium and silicon. For example, in n-type germanium crystals the capture of
holes, particularly prominent at low temperatures (from 200°K
down), may be due to structural defects produced by electron or
neutron bombardment [46] or due to impurity atoms of copper [47].
The position of the capture level and the hole capture cross section
were determined for the latter case. It was found that each atom
of copper has one capture level lying at a distance slightly more
than 0.2 eV from the valence band. The hole capture cross section was found to be independent of temperature. The position of
the trap level and the hole capture cross section ($S_p = 1.5 \times 10^{-16}$
cm^2) were practically identical with the results of Burton et al.
[35], who determined, in n-type Ge, the parameters of copper
atoms acting as recombination centers at room temperature (see
above); according to Burton et al., $S_p = 1 \times 10^{-16}$ cm^2. Independent experiments in which the temperature dependence of the
electron capture cross section was found for strongly doped p-type
germanium showed that the electron capture cross section S_n by
the same level of copper in Ge decreased steeply on increase of
temperature.

The data just presented show that the acceptor level of copper
in germanium, which acts as a recombination level at room temperature, becomes a hole trap at low temperatures due to the temperature dependence of the quantity S_n. It is very likely that other
impurity centers may, depending on the conditions in the crystal
(in particular, on temperature) act as recombination centers or
as minority-carrier traps.

The deep levels of Fe, Co, and Ni act precisely in this way
in germanium [48] (the investigation of these levels is not yet
complete). Haynes and Hornbeck [45] established that several
types of trap with different parameters exist in silicon single crystals grown from the melt (in a quartz crucible) by the uniform
pulling of a rotating single-crystal seed. Further studies showed
that in crystals grown without rotation the trap concentrations
were at least one order of magnitude lower. Moreover, it was
found that there was a direct relationship between the content of
oxygen dissolved in Si (cf. Chapt. I) and the trap concentration.
However, the majority of oxygen atoms in the interior of silicon
crystals were electrically inactive under normal conditions and

the traps consisted of more complex entities in which the oxygen atom was one of the constituents [49].

F. Recombination and Capture of Carriers by Surface Centers in Semiconductors

The surface of a semiconductor is an unavoidable macroscopic departure from the crystal lattice periodicity. Electron processes at and close to the surface of a semiconductor are affected not only by the capture and recombination centers but also by the state of the space charge region which is due to the existence of local centers at the surface. Such surface centers may be directly due to the termination of the lattice periodicity (Tamm levels) or due to adsorbed atoms or molecules. The investigation of surface processes is a special independent branch of the physics and chemistry of semiconductors in which considerable success has been achieved in recent years, particularly in the case of germanium single crystals. *

The recombination and capture of carriers by centers in the space charge layer near the surface frequently strongly affect the steady-state value and kinetics of the photoconductivity. Recombination through capture by surface centers may considerably alter the spectral dependence of the photoconductivity in the fundamental absorption band.

One of the simplest and frequently used methods of eliminating surface effects is the measurement of photoconductivity using semiconductor samples in the form of plates of various thicknesses. In this case, one can frequently determine the surface recombination velocity by using sufficiently weakly absorbed light. In a very thin plate of thickness d, which is small compared with $1/\alpha$, the measured "effective" lifetime τ_{eff} is related to the lifetime τ, representing the interior of the material, and to the surface recombination velocity s by

$$\frac{1}{\tau_{eff}} = \frac{1}{\tau} + \frac{2s}{d} \tag{2.51}$$

provided the excitation is weak (the density of excess minority carriers much smaller than the equilibrium density). Quantitative data on the relationship between the probabilities of volume

* See, for example, the review by Low [50] and the earlier review of Brattain and Garrett [51].

and surface recombination can be obtained also by analyzing the shape of the spectral dependences of the photocurrent which is related to the number of light-generated nonequilibrium carriers. A typical curve has a region where the photocurrent is constant due to the strong absorption within the fundamental band; on increase of wavelength, the photocurrent increases because the pair generation occurs not only at the surface but also in the interior of the crystal. The case of a photocurrent in a crystal with a p-n junction was discussed above.

We note that the concept of the surface recombination velocity s as a characteristic constant should be used with caution in quantitative analysis of photoelectric phenomena in semiconductors because in many cases the value of s depends strongly on the excitation intensity.

§ 13.   Dependence of the Photoconductivity on the Excitation Intensity Quenching of the Photoconductivity Negative Photoconductivity

We shall consider a simple model of a semiconductor free of traps: under steady-state conditions, the rate of generation of free carriers is equal to the recombination velocity. In the absence of radiative excitation

$$\sum_i g_i = n_0 v \sum_i S_i N_i, \tag{2.52}$$

where n_0 is the equilibrium (dark) electron density; g_i is the rate of thermal liberation from centers of type i; S_i and N_i are, respectively, the cross section for recombination capture and the concentration of centers of type i.

If an excess electron density Δn is established by optical excitation (generation rate g_r), then

$$g_r + \sum_i g_i = (n_0 + \Delta n) v \sum_i S_i N_i. \tag{2.53}$$

We shall assume that: (1) there is only one type of recombination center; and (2) both thermal and optical liberation from these centers is possible; then

$$g_r + g = (n_0 + \Delta n) vSN \tag{2.54}$$

and, since under these conditions $N = n_0 + \Delta n$,

$$g_r + g = (n_0 + \Delta n)^2 vS. \tag{2.55}$$

We shall denote by τ' the ratio

$$\frac{n_0 + \Delta n}{g_r + g} .$$

Then from the above expression it follows that:

$$n_0 + \Delta n = \left(\frac{g_r + g}{vS}\right)^{1/2}, \tag{2.56}$$

$$\tau' = [vS(n_0 + \Delta n)]^{-1}. \tag{2.57}$$

The meaning of these equations is simple for (a) an insulator, when $n_0 \ll \Delta n$ and $g \ll g_r$, and (b) a semiconductor with $n_0 \gg \Delta n$. For an insulator (with a wide forbidden band),

$$\Delta n = \left(\frac{g_r}{vS}\right)^{1/2}, \quad \tau' = (vS\,\Delta n)^{-1} = (vSg_r)^{-1/2}, \tag{2.58}$$

i.e., the value of τ' corresponding to the microscopic "lifetime,"* is inversely proportional to the photocurrent, and the photocurrent is proportional to the square root of the excitation intensity. This case represents, therefore, bimolecular recombination.

Usually, the experimental results cannot be described by this simple model which ignores trapping. The photocurrent rise, proportional to the square root of the excitation intensity, is observed relatively rarely; more frequent are the cases when the photocurrent is proportional to a fractional power of the excitation intensity, the power exponent being between 1/2 and 1; over a certain range of the intensities, a superlinear dependence is sometimes observed, with the power exponent greater than unity. For a semiconductor, it follows from Eqs. (2.56) and (2.57) [neglecting terms containing $(\Delta n)^2$]

$$\Delta n = \frac{g_r}{2n_0 vS}, \tag{2.59}$$

*See, for example, [61].

$$\tau' = (n_0 v S)^{-1}, \qquad (2.60)$$

i.e., the photocurrent is proportional to the excitation intensity and the nonequilibrium carrier lifetime is independent of this intensity.

When the density of nonequilibrium carriers is considerably greater than the initial density at thermal equilibrium, substantial departures from the linear increase of the photoconductivity with the excitation intensity may arise because of a change in the degree of the population of capture levels. If the concentration of the local centers, the positions of their levels in the forbidden band, and their carrier-capture cross sections are known, then the dependence of the probability of recombination on the excitation intensity and, secondly, the dependence of the photoconductivity on the excitation level can be calculated.

At present it is more usual for the experimentalists to solve the converse problem, i.e., to determine the parameters of local capture centers from the data on the carrier lifetime at various excitation levels or various temperatures. In many cases – for example, germanium and silicon – the analyses of the experimental data using the Hall-Shockley-Read statistics, and its generalization to several levels, have been very successful. However, the final identification of the dominant type of local center usually requires independent experiments additional to the lifetime determination. The nature and properties of recombination centers in other semiconductors, for example, substances with relatively wide forbidden bands (frequently used as photoresistors) have been studied less extensively, and the features of their photoconductivity are usually analyzed qualitatively using the method developed mainly by Rose [52, 53].

Rose's analysis is based on the concept of "demarcation lines (levels)" which separate certain types of level in the forbidden band. If the distribution of centers with levels having various energies in the forbidden band is sufficiently uniform, i.e., if the concentration of these centers varies more slowly than the function $\exp(-E_k/kT)$, where E_k is the energy of the k-th level, then according to Rose a demarcation line for electrons should be drawn where the probability of the thermal ejection of electrons to the conduction band is equal to the probability of hole capture. Similarly, a demarcation line for holes is drawn where the probabilities of excitation to the valence band and of capturing free electrons are equal.

In this case, the recombination velocity is governed by local levels between the two demarcation lines. The levels outside the region enclosed by the demarcation lines have very little influence on the recombination velocity.

The positions of the demarcation lines usually coincide almost exactly with the "steady-state Fermi levels." The mathematical concept of "steady-state Fermi levels" or "quasi-Fermi levels," is convenient in the discussion of steady-state nonequilibrium conditions in a semiconductor; the terminology follows from the analogy with the equilibrium Fermi distribution. Under thermal equilibrium conditions, the electron density in the conduction band is given by

$$n_0 = N_c e^{-\frac{E_F}{kT}}, \qquad (2.61)$$

where N_c is the density of states, E_F is the gap between the bottom of the conduction band and the Fermi level. Substituting into this expression the steady-state value of the electron density under nonequilibrium conditions, n_c we define quantity E_F', i.e., the gap between the bottom of the conduction band and the "quasi-Fermi level" E_F':

$$n_c = N_c e^{-\frac{E_F'}{kT}}. \qquad (2.62)$$

Thus recombination levels are those levels which lie between the quasi-Fermi levels, the positions of which in turn determine the electron and hole densities in the bands. From this point of view, the difference between traps and recombination centers is of a statistical nature.

When the rate of generation of nonequilibrium carriers increases, their densities in the bands increase and the quasi-Fermi levels are displaced toward the band edges. Consequently, the levels which have acted as traps under weak excitation now become recombination centers. Conversely, when the generation of nonequilibrium carriers stops, their density decreases and the quasi-Fermi levels approach one another coming closer to the equilibrium position. Then the levels which have acted as recombination levels under intense excitation conditions become traps.

A. Rose [52], and R. Bube in his book [53] have discussed several cases when the dependence of the photoconductivity on the

excitation intensity can be satisfactorily described by suitably selecting the local center parameters in semiconductors with relatively wide forbidden bands.

It should be noted that one of the important assumptions made
by Rose – that the same centers may act both as traps and as recombination centers – is not a universal rule, since local centers
may sometimes act only as traps, i.e., they may have very low
sections for carriers of one sign.

Phenomena such as sensitization (increase of the photosensitivity of a semiconductor by introducing impurities), super-
linearity, and "quenching" of the photoconductivity are explained
qualitatively using a model including at least two types of local
center.

In the case of weak excitation, when the free carrier densities are low compared with the densities of bound electrons and
holes, the total densities of bound electrons and holes remain
practically constant. However, the localized electrons and holes
are redistributed between recombination and trapping levels by
intermediate transitions to the bands, and from the bands to the
levels. A mathematical analysis using this model is possible but
the final formulas are very cumbersome and they do not give a
clear picture [2]. We shall restrict ourselves to a qualitative
discussion of several cases.

A. Sensitization of Photoactive Semiconductors

Let us assume that levels of type 1 in Fig. 39 are responsible
for the short lifetime of free carriers. This means that the photoconductor has low sensitivity and after illumination its resistance
quickly returns to its initial dark value.

The introduction of centers of type 2 may lower the recombination velocity and increase the lifetime of carriers of one type.

To make the case definite, we shall assume that the four
values of the densities of carriers bound to local levels of types 1
and 2 (denoted, respectively, by n_{g1}, p_{g1}, n_{g2}, and p_{g2}) are very
roughly equal, except that n_{g2} is about twice as high as the other
densities. Let the cross sections for the capture of electrons
S_{n1} and holes S_{p1} by centers of type 1 be equal (10^{-15} cm^2) while
the centers of type 2 have the same hole capture cross section
($S_{p2} = 10^{-15}$ cm^2) but a much smaller electron capture cross section ($S_{n2} = 10^{-20}$ cm^2). In the absence of centers of type 2, the

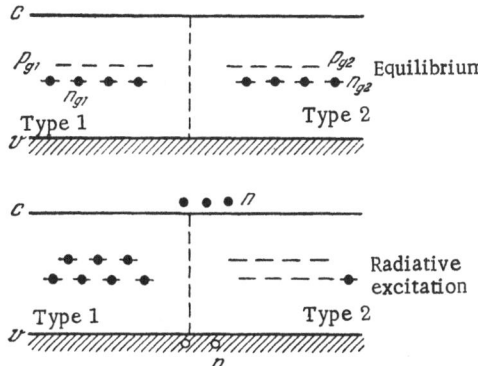

Fig. 39. Energy system explaining the mechanism of
sensitization of a photoconductor by the introduction
of impurity centers with deep levels. Centers of types
1 and 2 are separated in the figure although in fact
these centers are distributed at random in the crystal.

lifetime of nonequilibrium carriers τ is $(vS_{n1}n_{g1})^{-1}$, where S_{n1}
$= 10^{-15}$ cm^2; then, if $n_{g1} = 10^{15}$ cm^{-3}, the lifetime τ is the same
$(10^{-7}$ sec) for electrons and holes.

The introduction of centers of type 2 has the following effect:
holes are captured by local levels of these centers, which have
a low probability of electron capture. The electron recombination velocity decreases because of the small effective cross section for electron capture by centers of type 2, and because of a
redistribution of holes (with which electrons should recombine)
between levels of various types, which reduces the density of holes
at levels of type 1. The lifetime of a free electron then becomes
$(vn_{g2}S_{n2})^{-1} \approx 10^{-2}$ sec, because $S_{n2} = 10^{-20}$ cm^2. The average hole
lifetime decreases somewhat due to the increase of the concentra-
tion of capture centers but the final result is manifest as an increase of the photosensitivity (and of the slowness of the response)
by a factor of about 10^5.

In the example given, the parameters of the injected centers
were selected so that their appearance caused a strong rise of the
photosensitivity. It must be understood that it may frequently
happen that the lifetimes of both types of carrier are reduced. The
most important conclusion following from the example taken from
Rose's work is the fact that a redistribution of carriers between

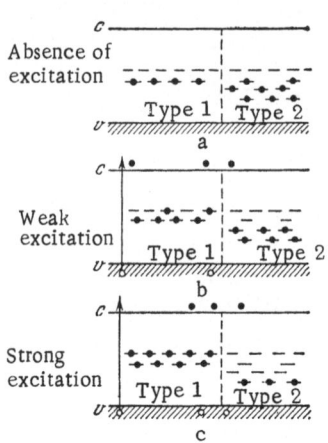

Fig. 40. Appearance of "superlinear" dependence of the photocurrent on the excitation intensity, according to A. Rose [52].

two types of local center may increase the photocurrent by many orders of magnitude.

Moreover, the above interpretation of the process of sensitization of photoconductors allows us to suggest that the "activating centers" introduced additionally are not necessarily the "source" of photoconductivity, since the initial photoionization process may not be associated with these centers.

Sensitization by introducing impurity centers with which the absorbed photons interact directly (typically, the introduction of gold atoms into germanium) usually gives rise to an additional spectral region of photosensitivity. However, the introduction into germanium of gold and other centers with deep levels (iron, cobalt, nickel, zinc) is also accompanied by sensitization, i.e., by an increase of the steady-state photocurrents and of the slowness of the response (cf., for example, [59, 36]).

B. Superlinearity

Sometimes the experimentally observed "superlinear" dependence of photocurrent on the excitation intensity can be explained by using the foregoing model with two types of center, assuming additionally that the levels of type 2 centers are distributed over a certain range of energies. The level system in Fig. 40 shows that with increase of the excitation intensity the quasi-Fermi levels shift closer to the bands and enclose a large number of type 2 levels. Then an increasing number of electrons is transferred to type 1 levels, as a result of which the relative photocurrent increases due to the increase of the free-electron lifetime.

Figure 41 shows typical superlinear lux-ampere characteristics of cadmium sulfide containing chlorine (donor) and copper impurities [60].

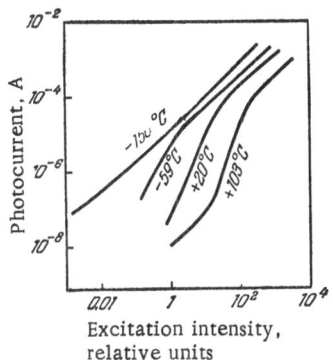

Excitation intensity,
relative units

Fig. 41. Dependence of the photocurrent on the excitation intensity for
CdS doped with chlorine and copper.

The discussion of the sensitization and of the superlinear
lux-ampere characteristics of a
semiconductor given above applies at relatively weak excitations, when the free-carrier densities are low compared with the
densities of bound electrons and
holes.

At high excitation intensities,
the free hole and electron densities are equal. The relative
times during which a given level
is occupied by an electron or a
hole, are $S_n (S_n + S_p)^{-1}$ and
$S_p (S_n + S_p)^{-1}$; they are inde-
pendent of the existence of other capture centers. The sensitivity of a photoconductor in the region of strong absorption cannot be
increased by the introduction of new capture centers. Therefore,
the superlinear parts of lux-ampere curves are always restricted
to the weak excitation region (Fig. 41).

C. Quenching of Photoconductivity

The experimentally observed reduction of a photocurrent,
excited by the absorption of light in one spectral region (usually
in the fundamental band region), by the additional illumination
with light from another spectral region (usually longer wavelengths),
is known as quenching. This form of quenching is known as "optical. " A similar reduction of the photocurrent is also possible
when the temperature is increased above a certain limit (temperature quenching). The phenomenon of the quenching of the photoconductivity is due to the photoionization or thermal ionization
of capture centers (traps) containing minority carriers, which increases the number of transitions to the recombination centers
and thus reduces the majority carrier lifetime. The optical quenching of the photoconductivity by additional illumination in the infrared region is observed, in particular, in germanium doped
with gold, iron, nickel, and other impurities, forming deep levels.
In germanium crystals containing these impurities, it has been
possible to identify the trapping levels which are active in the en-

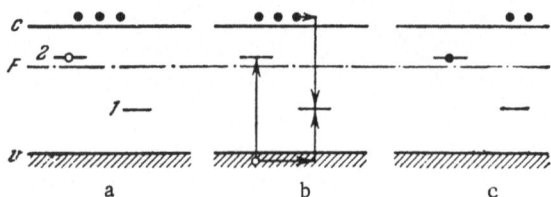

Fig. 42. Electron transitions in the case of negative photoconductivity.

hancement of the photosensitivity and in slowing down the response,
as well as those involved in the infrared and temperature quenching [59].

D.   Negative Photoconductivity

The absorption of radiation is sometimes accompanied not
by a rise but by a reduction of the electrical conductivity. This
phenomenon is known as the negative photoconductivity. In contrast to the optical quenching, which reduces the photoconductivity,
the negative photoconductivity reduces the electrical conductivity
below the thermal equilibrium level. This effect is observed relatively rarely and has not been investigated systematically. To
explain the mechanism of the negative photoconductivity, various
suggestions have been put forward. Borshchevskii [54] assumed
that this phenomenon is due to an illumination-induced increase of
the polarized region of a crystal, associated with defects and impurities, which lowers the resultant electrical conductivity of a
sample to which an external field is applied.

Miselyuk [55], who investigated the photoconductivity of Ag_2S,
established that the negative photoconductivity appeared on the
addition of 1-2% PbS and concluded that the illumination "activated"
the levels so that they were capable of capturing electrons from
the conduction band. Borisov and Kanev [56] investigated zinc
oxide (ZnO) and concluded that the light generates excitons which
are scattered on thermally ionized atoms of excess zinc and thus
transfer electrons from the valence band to the levels of the excess zinc. As a result of this, holes are generated, which increases the velocity of recombination of free electrons and reduces the electrical conductivity below the dark value. A detailed
interpretation of the negative photoconductivity of germanium

containing impurities with deep levels representing various charged states of impurity centers, was given by Stockmann [57], who concluded that the recombination velocity of majority carriers increases and the electrical conductivity decreases when minority carriers are generated by the photoionization of impurity centers.

Electron transitions in the case of the negative photoconductivity (Fig. 42) are assumed to occur under the following conditions: (a) the rate of thermal ionization of levels 2 is less than the velocity of recombination of electrons and holes at levels of type 1;(b) holes do not recombine with electrons at levels 2; (c) levels 2 lie above the Fermi level; (d) the cross section for the capture of majority carriers by centers 2 is much smaller than for centers 1;(e) the concentration of centers 1 and their cross section for the capture of minority carriers is not too small. The small likelihood of satisfying simultaneously all these conditions is, according to Stockmann [57], responsible for the fact that the negative photoconductivity is relatively rare.

Recently, Dobrego and Ryvkin [62], who detected negative photoconductivity at helium temperatures in germanium relatively heavily doped with elements of groups V or III, explained the effect by assuming that an increase of the population of donor levels on illumination reduces the electrical conductivity which, under experimental conditions, involves electron "jumps" from neutral to charged donors.

Chapter III

IONIZATION OF SEMICONDUCTORS BY CHARGED HIGH-ENERGY PARTICLES

§14. Nature of the Interaction of Charged High-Energy Particles with Matter Ionization Energy Losses

It is well known that the energy losses accompanying the passage of charged high-energy particles through matter are almost completely due to the excitation of bound electrons. These energy losses determine the particle range in a given substance and are known as the ionization losses, although in fact not all the electron transitions are accompanied by the ionization, because excitation is possible. When fast electrons pass through matter, some of the energy is converted into electromagnetic bremsstrahlung radiation. At high energies exceeding a certain critical value E_C, the bremsstrahlung losses in the electric field of nuclei may be greater than the ionization losses. According to Bethe and Heitler, the ratio of the radiation to ionization losses for fast electrons is given by

$$A = \frac{\left(\frac{dE}{dx}\right)_{\text{rad}}}{\left(\frac{dE}{dx}\right)_{\text{ioniz}}} \approx \frac{E_0 Z}{1600 mc^2},\tag{3.1}$$

where E_0 is the initial energy of the particle, and Z is the nuclear charge. It follows that the critical value E_C, corresponding to $A = 1$, and equal to $1600\ mc^2/Z$, amounts to about 11 MeV for germanium and 28 MeV for silicon.

Calculation of the average ionization losses and ranges of charged particles, and comparison of the results obtained with experimental data are important problems in nuclear physics (for a detailed discussion see, for example, [1]).

From the point of view of the physics of high-energy particles, the specific loss of particle energy (the rate of loss of energy), i.e., the quantity –dE/dx, is of special interest, and the
determination of this quantity has been discussed in various theoretical papers, including the classical work of N. Bohr, K. Muller,
H. Bethe and others [2, 3]. However, these papers, and numerous
experimental investigations of the passage of fast charged particles through matter, did not give direct information on the change
of state of the electrons in the matter to which the particle energy
was transferred. In the theory of ionization losses, developed in
its modern form by H. Bethe, the binding energy of electrons in
the matter to which the energy is transferred enters the formulas
describing the ionization losses only as the "average ionization
potential" or "average energy of electron excitation" I.

According to Bethe, the specific energy loss –dE/dx for relativistic electrons is

$$-\frac{dE}{dx} = \frac{2\pi N q^4}{m v^2} Z \left[\ln \frac{m v^2 E}{2 I^2 (1 - \beta^2)} \right.$$

$$\left. - (2\sqrt{1 - \beta^2} - 1 + \beta^2) \ln 2 + 1 - \beta^2 + \frac{1}{8} (1 - \sqrt{1 - \beta^2})^2 \right] + \ldots, \quad (3.2)$$

where E is the kinetic energy of a fast electron, v is its velocity,
$\beta = v/c$ (c is the velocity of light), and N is the number of atoms
in 1 cm^3 of the matter through which the electron is traveling. For
small values of β, the above formula reduces to

$$-\frac{dE}{dx} = \frac{4\pi q^4 N}{m v^2} Z \ln \frac{m v^2}{2 I} \sqrt{\frac{q}{2}} . \quad (3.3)$$

The expressions (3.2) and (3.3), quoted here as examples, as
well as similar formulas for heavy charged particles, agree satisfactorily with the experimental data on the energy losses in
gases and thin metal foils, if it is assumed that for light elements
with Z < 15 we have I ≈ 11.5Z eV, and for heavier nuclei I ≈ 9Z
eV. The problem whether the state of the matter traversed by a
charged particle affects the specific ionization losses has not
been solved yet. However, from the theoretical point of view,
which regards ionization as the result of the interaction between
the fields of the traveling particle and of the electrons in the mat-

ter, we can expect the following features: the same substance (for example, boron) should retard charged particles in a different way in the solid (crystalline) state from in the free atomic state. This is because the valence electrons in a crystal are able to cross over to an unfilled band, whereas there is no such possibility in an assembly of free atoms. In matter consisting of heavy nuclei, the effect on the outer electrons, which are affected by chemical binding, should be weak.

In addition to the bremsstrahlung energy losses, which are important only at very high electron energies, there are other physical processes which accompany the passage of high-energy particles. One of these is the direct excitation of crystal lattice vibrations. F. Seitz showed that the energy losses in the latter process are negligibly small compared with the ionization losses [4].

The origin of the "characteristic" energy losses is not very clear: the effect consists of the conversion of the initially mono-energetic fast-electron beam into a beam with an energy spectrum having several bands or lines which are characteristic of the substance traversed by the electrons. It is very likely that in some cases the characteristic losses are directly related to electron processes involving weakly bound electrons in the outer shells of atoms [5]. Thus, experimental investigations of the characteristic losses in semiconductors – which have only recently begun – may yield results important not only in the physics of semiconductors but also in the theory of ionization losses. The current theories of the ionization losses by fast electrons do not allow for the characteristic losses.

The direct transfer of energy to atoms in matter, which under certain conditions produces structural defects (radiation defects), will be discussed in greater detail in Chapt. V. In the case of electron bombardment, the fraction of the energy lost by knocking out atoms is only a small portion of the total initial electron energy.

In the case of heavy-particle bombardment, it is assumed – following F. Seitz [4] – that elastic collisions with atoms begin to predominate over electron excitation when the kinetic energy of a particle E is reduced to

$$E_i = \frac{E_g}{8}\frac{M_A}{m},\tag{3.4}$$

where M_A is the mass of an atom of the matter being traversed, m is the electron mass and E_g is the forbidden bandwidth. For example, in the case of germanium $E_i \approx 12,000$ eV, and therefore a 5 MeV alpha particle loses about 0.2% of its energy by elastic collisions in a germanium crystal. The phenomenon of electron capture by heavy particles at the end of their path introduces indeterminacy in this approximation. The experimental data obtained with Wilson chambers and nuclear emulsions show that the phenomenon of charge exchange in the case of alpha particles markedly affects the nature of the ionization at the end of their path. However, an increase in the number of elastic collisions affects mainly the number of radiation defects, while the fraction of the energy lost in elastic collisions remains relatively small even in the case of protons and alpha particles with energies of the order of 10^6 eV.

In concluding our discussion of this problem, let us consider the particle range and the spatial distribution of the ionization. A knowledge of the particle range and the nature of the changes in the trajectory due to the interaction with matter, as well as the data on the initial distribution of nonequilibrium carriers generated in the interior of a semiconductor crystal subjected to particle bombardment, are necessary not only in practical applications (for example, semiconductor particle counters) but also in the design of experiments for investigating ionization by radiation.

In experiments with heavy particles, we can use the reliable data on the particle range accumulated in numerous papers on experimental nuclear physics (see, for example, [1]). An important aspect, which facilitates the interpretation of the results, is the low probability of the scattering of heavy particles along directions making large angles with the initial path of the particle. We may assume that the direction of particle motion remains practically unaltered as long as the ionization losses predominate. The ionization region near the trajectory of a heavy charged particle in a crystal is, in the first approximation, enclosed by a cylinder whose axis coincides with the particle trajectory. This approximation ignores the fraction of the energy transferred to fast secondary electrons (δ electrons) and the possibility of the migration of the excitation energy by emission and the re-absorption of x-ray quanta which are generated when strongly bound electrons of the inner shells are knocked out. According to the data obtained

with a Wilson chamber for protons of up to 10 MeV energy [1],
several δ electrons may appear in 1 cm of the proton path in a
gas at normal pressure; the maximum energy of these electrons
corresponds to double the velocity of the heavy particle and is

$$E_\delta = 4E \frac{m}{M},$$

where M is the mass of the particle; for protons of 10 MeV energy, E_δ reaches 20 keV. Delta electrons may cause ionization far beyond the initial track (the trajectory region) of the particle, and in some cases may affect the quantitative results.

In experiments with fast electrons, it is difficult to obtain
data on the ionization distribution in matter because of the scattering of incident electrons by atoms. The range of an electron
is much greater (because of its small mass) than that of a heavy
particle (for example, a proton) of the same energy, and the total
number of scattering acts accompanying a given energy loss is
considerably greater than that for heavy particles. In most cases
one deals with a beam of monoenergetic electrons incident normally on the flat surface of a sample. On entering the sample,
the electrons are scattered and the average distribution of the
ionization losses measured from the surface along the initial
beam direction differs strongly from the ionization distribution
for single electrons along its path. In analyzing experimental
data, special attention should be paid to the fact that some fast
electrons scattered at large angles leave the bombarded sample.
L. Spencer [6] developed a method for calculating the average energy losses for electron beams traversing matter. The distribution of the ionization losses of an electron beam in Ge and Si was
calculated, using the Spencer method, by B. Ya. Yurkov for several different values of the initial kinetic energies [7].

In connection with a study of the possibility of direct conversion of the energy of beta rays into electrical energy, data were
obtained at the P. N. Lebedev Physics Institute of the USSR Academy of Sciences on the distribution of the specific ionization in
germanium crystals bombarded with monoenergetic 920 keV electrons. Measurements were carried out using ionization chambers
and narrow air gaps between two germanium plates. The curves
given in Fig. 43 represent the change in the ionization losses with
increasing thickness of germanium plates bombarded with a parallel beam of electrons [8].

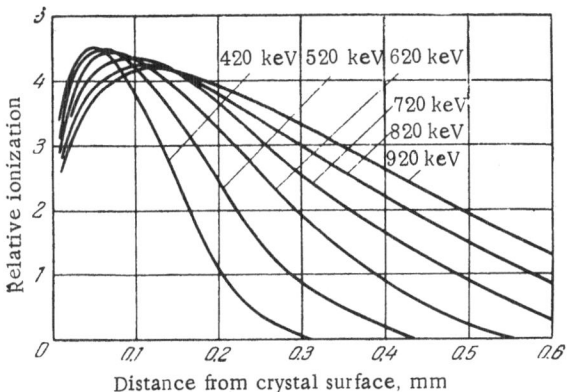

Fig. 43. Distribution of ionization caused by fast electrons
in a germanium crystal [8].

The results of Yurkov's calculations, carried out on the assumption that the initial energy of the generated nonequilibrium
electrons and holes is low (not more than several electron volts),
were in good agreement with the experimental data.

§ 15. Experimental Determination of the Average Energy of Formation of a Nonequilibrium Carrier Pair in Semiconductors

In the case of ionization in gases, which has been studied very
extensively, the total number N of ion pairs produced during the
slowing down of a particle is proportional to the reduction of its
kinetic energy, i.e., $N = \Delta E / \varepsilon$, where ε is the average energy
needed to form an ion pair.

Numerous experiments on the total ionization in gases showed
that the value of ε is close to 30 eV and, within the limits of the
experimental error, is independent of the mass, charge, and velocity of the fast particle causing the ionization. Moreover, the
value of ε shows no direct relationship with the ionization potential of the gas [1].

The value of ε for gases is important since it governs the
feasibility and sensitivity of gas-filled particle counters and ionization chambers.

The analogous quantity in the case of semiconductor devices
used in the detection and spectrometry of fast particles, and in

semiconducting converters of nuclear radiation energy, is the average energy required for the formation of a carrier pair: an electron and a hole. This quantity is also denoted by ε .

As mentioned above, the "ionization losses" of a fast charged particle are due to the interaction of this particle both with valence electrons and with electrons of the inner shells, which are relatively tightly bound to their atoms. In Bethe's theory, this interaction is reflected by the "average ionization potential" I. Inner secondary electrons of sufficiently high energy give rise to tertiary electrons, etc.

The excitation energy of the inner atomic shells is converted into the energy of relatively soft x rays which are normally absorbed within the crystal, producing nonequilibrium carrier pairs at the expense of some of their energy.

The experimental study of the ionization in a solid caused by charged particles was initially started with the idea of producing " solid ionization chambers" and counters which should have many advantages compared with the gas-ionization devices used for particle detection and energy determination. The first crystals to be investigated were alkali halides and diamond.*

Estimates of the average ionization energies given in earlier work (approximately up to 1950) are very contradictory, mainly because of the unavailability of perfect crystals with reproducible properties, and because of difficulties in the elimination of the volume polarization effects.

Considerable success in the investigation of the ionization in semiconductors followed with the preparation of perfect single crystals with p-n junctions and the development of the theory of nonequilibrium processes in such crystals [11].

We shall consider next the experiments which gave the quantitative data on ionization by alpha-particles and fast electrons in Ge and Si single crystals, together with some recent data on the ionization of diamond by electrons.

A. Germanium

McKay [12] investigated ionization by alpha particles in the strong-field region of a p-n junction. We shall consider a semi-

---

* A detailed description of the work up to 1949 can be found in the reviews of A. Chynoweth and R. Hofstadter [9, 10].

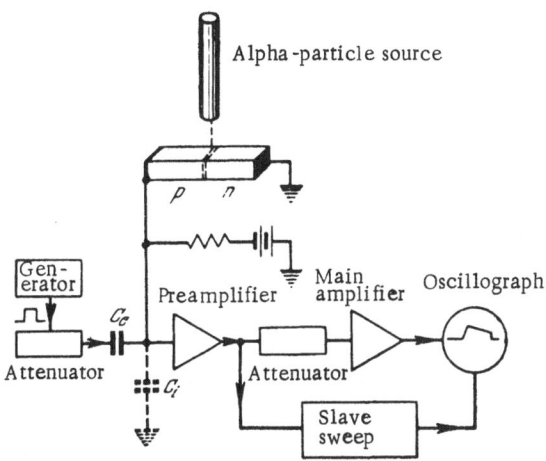

Fig. 44. Experimental layout used to determine the average energy of ionization by alpha particles in semiconductor crystals with p-n junctions.

conductor slab (Fig. 44), one half of which is p-type and the other n-type. If a potential, which is positive with respect to the p-type region (reverse bias), is applied to the n-type region, then the junction resistance is high and practically the whole voltage drop is concentrated at the junction. The carrier-depleted strong-field region becomes analogous to an insulator separating two conductors. If the bombarding particles are slowed down in the strong-field region, then the nonequilibrium electrons and holes generated by ionization move rapidly to the boundaries of the junction. McKay carried out control tests, consisting of optical excitation with a monochromatic light source of known intensity, and showed that the carriers representing pairs generated in the junction region are completely separated, i.e., there is no recombination of capture of carriers and the quantum yield is unity. *

Figure 45a shows the equivalent circuit of a germanium crystal and the amplifier input. R_b and C_b are the parameters of the junction, R_i and C_i are the components of the amplifier input impedance, and R_s is the series-connected resistance of the interior of germanium. On the application of a reverse voltage to

* In strong fields in the pre-breakdown region, impact multiplication of carriers is possible at the expense of the energy acquired during motion in the field [13]. This effect did not occur in the experiments described above.

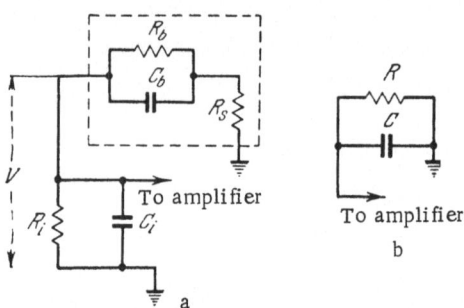

Fig. 45. Equivalent circuit of a crystal with a p-n
junction and of the amplifier input.

the junction, $R_b \gg R_s$ and, consequently, we can neglect R_s. Then the equivalent circuit reduces to a simple parallel RC circuit (Fig. 45b) where $C = C_b + C_i$ and $R = R_b R_i (R_b + R_i)^{-1}$. The carrier transit time across the strong-field region is considerably shorter than the RC relaxation time, so that the result of ionization by a particle is equivalent to the formation of a pulse of current through the barrier (blocking) layer, the resultant voltage $V = QC^{-1}$ being maximum at the barrier (Q is the effective charge transported across the strong-field region).

Germanium samples used to study the alpha particle ionization were single-crystal slabs 2 cm long and 1×1 mm^2 in cross section with a p-n junction located approximately in the middle. The ends of the crystals were sand-blasted and coated with rhodium to obtain nonrectifying low-resistance contacts. Probe measurements along the sample showed that under a reverse bias of several volts the voltage drop at the contacts and across the uniform parts of the crystal amounted to a fraction of 1% of the voltage drop across the junction.

Figure 44 shows the experimental setup. One end of the sample was grounded, and the other connected to the input of the wideband amplifier. The voltage was applied to the sample through a shunting resistance approximately equal to the barrier resistance. The output signal of the preamplifier was used also to trigger the driven sweep of the oscillograph. The attenuator connected between the preamplifier and the main amplifier included a 0.17 μsec delay line (a coaxial cable). The maximum bandwidth of the amplification channel was 35 Mc, and the maximum am-

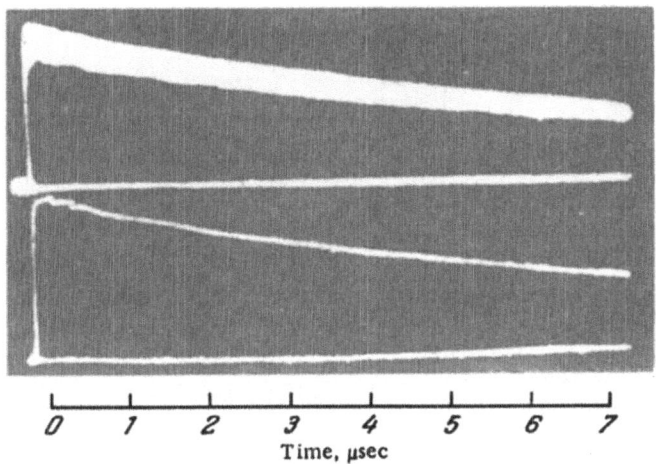

Fig. 46. Pulses representing ionization by alpha-particles near a p-n junction in a germanium crystal [12].

plification was 100 dB. The calibrating circuit included a square-pulse generator (0.01 μsec rise time). The generator output was connected, through the attenuator and the small capacitor C_C, to the preamplifier input. The calibrating pulse, reaching the capacitor C_C, formed a current pulse (or, more correctly, a charge pulse) at the input of the preamplifier. Since the resistance R_S was negligibly small (Fig. 45), this was equivalent to the formation of a current pulse across the p-n junction during the passage of an alpha particle. Comparing the oscillograms of the calibrating pulses and the pulses produced by alpha particles, one could determine the charge transported across the junction by electrons and holes generated by a single alpha-particle.

McKay used alpha particles from a polonium source fitted with a collimator. The maximum scatter of the points of incidence of the alpha particles on the crystal amounted to 3.5×10^{-3} cm, and was somewhat greater than the width of the strong-field region, which did not exceed 5×10^{-4} cm under a reverse voltage of several volts. Consequently, there was some scatter of the pulse amplitudes, since the diffusion of carriers generated outside the strong-field region, before their capture by the field, was accompanied by recombination. Calculations showed that 90% of the carriers generated at a distance of 10^{-3} cm from the

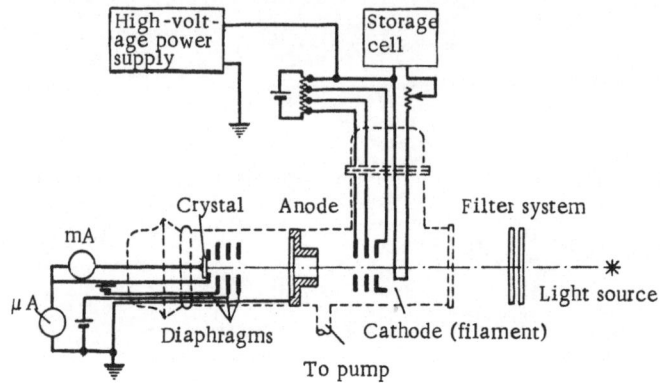

Fig. 47. Experimental layout for investigating the ionization by fast electrons in a semiconductor crystal with a p-n junction.

junction should have diffused to the junction in less than 0.5 μsec.
The measurements were carried out in vacuum in order to eliminate alpha-particle scattering and the ionization of air near the
sample.

Figure 46 shows a single pulse (below) and a train of pulses
(above) representing ionization by single alpha particles in a germanium crystal with a p-n junction. The observed scatter of the
pulse amplitudes is in agreement with the scatter calculated using
the known value of the alpha-particle beam width. The lower os-
cillogram represents a calibrating pulse. The maximum am-
plitude of the pulses due to alpha particles was used as the final
value in the calculations. The maximum pulse heights, obtained
for germanium samples of different resistivities, were equal to
within 5%. A 0.3-15 V change of the voltage across the junction
did not alter the maximum height of the pulses. The transit time
of electrons and holes across the barrier was not determined,
but it was not more than 0.02 μsec (the relaxation time of the
amplifier).

According to McKay, the carrier transit time across the
strong-field region was not more than 10^{-10} sec, but it was impossible to measure it. * The final value of the charge per single

* McKay assumed that the field acts on free electrons and holes immediately after
generation. The justification of this assumption could be disputed in the strong-
ionization region (the alpha-particle track), because initially this track is in the
plasma state.

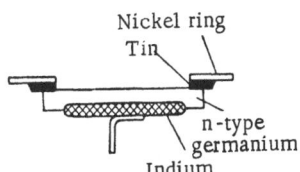

Fig. 48. Section through an n-type germanium crystal with a fused indium contact and a nonrectifying tin contact.

incident alpha particle, was, according to McKay, $Q = 1.77 \times 10^6$ q for germanium (q is the electronic charge).

To determine the average energy ε_α needed to produce a carrier pair, we must divide the initial energy of the alpha-particle, E_α, by Q/q. Thus, for $E_\alpha = 5.298$ MeV

$$\varepsilon_\alpha(\text{Ge}) = E_\alpha \frac{q}{Q} = 3.0 \pm 0.4 \text{ eV}.$$

Later, McKay and McAffe [13] carried out similar tests on silicon crystals with p-n junctions: they found that $\varepsilon_\alpha = 3.6 \pm 0.3$ eV for silicon.

Because of the considerable scatter in the fast-electron directions, it is difficult to use the above method for collecting quantitative data on ionization by electrons. Therefore, V. S. Vavilov, L. S. Smirnov, and V. M. Patskevich [14] used excitation outside the strong-field region in their investigation of the ionization of semiconductors by fast electrons. They had to allow for the recombination of carriers diffusing toward the p-n junction.

Germanium and silicon were bombarded with 5-25 KeV electrons in vacuum (Fig. 47). In view of certain features which were found during this work, we shall consider first the method used for germanium and then we shall move on to the silicon experiments.

The samples were in the form of slabs, 0.3-0.5 mm thick, cut from n-type germanium single crystals. Indium was fused to one of the surfaces of a slab and a nonrectifying ring contact of tin also was provided (Fig. 48). The side opposite to the indium contact was irradiated, the electrons being stopped completely in the n-type region. The method of determining the value of ε consisted of measuring the short-circuit current J_2 between the indium electrode and the nonrectifying contact, and the current J_1 representing the accelerated-electron beam incident on the crystal. The value of ε was proportional to the ratio of the currents: $\gamma_1 = J_2/J_1$.

The total number of carrier pairs, N_0, generated by ionization is higher than the flux of holes across the junction, $N_2 = J_2/q$. The ratio $N_2/N_0 = \beta$ is the quantum efficiency, which, for an elec-

tron penetration depth much smaller than the germanium slab
thickness, is independent of the electron energy [cf. Eq. (2.24)].
The value of the intrinsic "multiplication factor" γ is related to
β by $\gamma = \gamma_1 \beta^{-1}$. On the other hand,

$$\varepsilon = \frac{E}{\gamma}, \tag{3.5}$$

where E is the kinetic energy of fast electrons, measured in electron-volts. Thus, finally

$$\varepsilon = \frac{E}{\gamma} = \frac{\beta J_1 E}{J_2}. \tag{3.6}$$

Obviously, under the test conditions described above (surface excitation), the quantum efficiency β should be the same for any
nonequilibrium carrier irrespective of its mode of generation.
Nevertheless, it has been established that the value of β depends
very strongly on the surface state of germanium and changes
during pumping and also during and after electron bombardment.
This is due to large variations in the surface recombination velocity s. In order to avoid the necessity of determining s during
the experiment, the following method was used. During electron
bombardment the crystal surface was illuminated with monochromatic light of $\lambda = 1.05\,\mu$, which was absorbed in a layer 1 μ
deep. The intensity of the light beam was determined with a cal-
ibrated thermopile; the values of the reflectivity were measured
separately. Control tests showed that the coefficient β was independent of the wavelength of the illumination in the range λ
$= 0.8-1.5\,\mu$. As shown earlier (Chapt. II) the photoionization
quantum yield is unity for Ge if $\lambda = 1\,\mu$. Using the same instrument to measure the photocurrent J_{ph2} and the bombardment-in-
duced current J_2 (Fig. 47), one could determine β and calculate
the value of ε. The electron gun construction made it possible to
direct the monochromatic light beam along the electron beam.
Diaphragms were placed in front of the crystal to restrict the beam,
that nearest to the crystal being used to collect the secondary
electrons. Control measurements showed that at E = 5-15 keV the
secondary emission coefficient was 0.25-0.3. As is known, the
majority of secondary electrons has low energies. Thus, the
fraction of the primary beam energy taken away by secondary
electrons was negligible. Another small fraction of the primary-

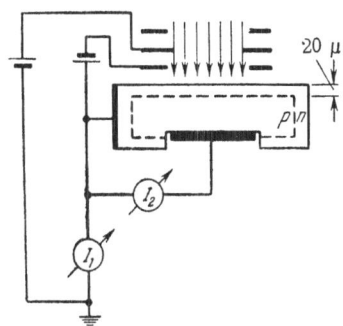

Fig. 49. Section through a silicon crystal with a p-n junction
and a circuit for measuring the
primary current and the current
between p- and n-type regions of
the crystal.

beam energy (less than 1%) was transformed into x radiation. To estimate the energy lost via secondary electrons and x rays, a calorimetric control test was carried out. The germanium crystal temperature rose by several degrees on the absorption of fast electrons. A similar temperature rise can be induced by passing a current through the crystal. Comparison of the energy necessary to heat the crystal by the Joule effect with the quantity $J_1 \cdot E$ showed that the energy lost by secondary electron emission and x-ray emission did not exceed 4%. The electron bombardment was applied to several crystals with different surface states (etching in hydrogen peroxide and CP-4 mixture). The values of β for the same crystal differed considerably; nevertheless the values of ε, calculated as outlined above, were practically identical. The variation of the pressure in the vacuum chamber containing the sample from 2×10^{-4} to 2×10^{-6} mm Hg, and the variation of the duration of bombardment (which lowered the quantum efficiency by a factor of several times) also failed to affect the value of ε, which was 3.6 ± 0.4 eV. Furthermore, ε was independent of the electron energy E within the range 5–15 keV.

Similar investigations were carried out at higher electron energies (E = 420 keV) using an electron accelerator, and with the bombarded germanium crystals placed in air. The accuracy of the measurement of ε was reduced by the fact that the depth of penetration of the electron beam was comparable with the crystal thickness. Since the intensity of light of any wavelength decreases exponentially with depth in a crystal, but the ionization curve (the distribution of the carrier generation with depth) is more complex and has a maximum, the method of comparing the optical and electron excitations cannot give the same accuracy which was obtained for electron beams of 5–15 keV energies. The value of ε was deduced from the experimental data allowing for the distribu-

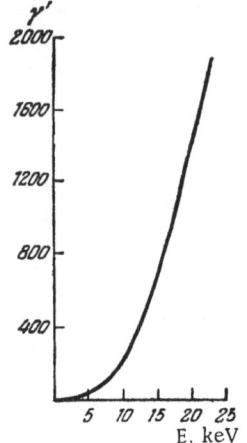

Fig. 50. Dependence of the carrier multiplication factor in silicon on the fast-electron energy E.

tion of the ionization in germanium with depth; it was 4.5 ± 1.5 eV for bombardment with 420 keV electrons.

Thus, within the limits of the experimental error, the values of ε for alpha particles and electrons were identical and, moreover, the average energy ε did not vary with the incident particle (electron) energy in the range 5-500 keV.

B. Silicon

To determine the average energy of the carrier-pair formation by fast electrons (10-25 keV) in silicon, we used the same "multiplication coefficient" method as before [15, 16]. In the apparatus shown in Fig. 47, we employed p-type single crystals with p-n junctions prepared by thermal diffusion of phosphorus from the gaseous phase. The method of preparing these junctions is described in [17]. An electron beam fell on the surface of the n-type region (Fig. 49). Because the carrier diffusion lengths were short in heat-treated silicon, we used crystals with p-n junctions lying about 20 μ below the bombarded surface. The quantum efficiency β for crystals in which excess carriers were excited by monochromatic light incident on the same surface as the electrons depended on the wavelength λ and, consequently, on the spatial distribution of the electron-hole pair generation. It was established that, in contrast to germanium, the quantum efficiency did not vary during pumping from atmospheric pressure to 10^{-7} mm Hg, or after bombardment with electrons of energies of the order of tens of keV. This can be explained by the existence on the surface of silicon of a thin but very stable and uniform oxide layer. This explanation is supported also by the fact that, in contrast to electron bombardment, ion bombardment lowered the value of γ' by increasing the surface recombination velocity. Figure 50 shows a typical dependence of the electron multiplication coefficient γ' of silicon (without allowance for recombination losses) on the electron energy E up to 25 keV. The shape of the curve can be explained in a natural way by the existence of a "dead" (inactive) layer in which the electron energy is absorbed without generating nonequilibrium carriers.

The value of ε was determined for each value of the electron energy by comparing the multiplication coefficient γ' for this energy with the value of the quantum efficiency β for light of such wavelength that the reciprocal of the absorption coefficient α_λ^{-1} was equal to the average depth of penetration of the electron beam of given energy E. To estimate the distribution of the ionization energy losses with depth, the calculations of B. Ya. Yurkov, who allowed for the scatter of the electron directions, were used. The ratio β/γ', proportional to ε, should be independent of the thickness of the layer in which the ionization occurs. Starting from his assumption, we estimated, from the $\gamma = \gamma(E)$ curve, the "dead" layer thickness, which was found to be $0.5-0.7\ \mu$; the average value of ε, determined by this method, was 4.2 ± 0.6 eV.

C. Beta-Particle Ionization of Diamond

At present, diamond is not widely used as a semiconducting material. However, there are grounds for expecting that in many special cases (strong fields, high temperatures, etc.) it may become very valuable. Comparison of the data on ionization in diamond with the data on silicon and germanium is particularly interesting because the crystal structure of each of these elements is the same and one would expect similar behavior with regard to the radiation ionization because the latter is governed primarily by the energy band structure and the lattice vibration spectrum.

In this connection, we shall consider briefly the results of the recent work of P. Kennedy [18], which differed considerably from the earlier experiments of McKay et al. [19, 9] who reported a value of ε close to 10 eV, which was much too low.

In contrast to the earlier work on crystal counters [9, 10], Kennedy used monoenergetic beta particles, selected with a magnetic spectrograph. Kennedy made sure that he reached the saturation conditions for the pulse heights in a strong electric field applied to uniform samples of natural diamond. To allow for the fraction of the beta-particle energy lost by scattering and the emission of x rays, the following procedure was adopted. The sample was surrounded almost completely – with the exception of a narrow channel through which passed the beta-particle beam – by

a plastic material incorporating a scintillator. Each scintillation flash in the plastic was recorded by a photomultiplier connected in a circuit ensuring coincidence with the conduction pulses
in the diamond. In analyzing the results to determine the average
ionization energy ε, Kennedy selected those conduction pulses
which represented complete absorption of the energy of the incident beta particle, i.e., those which were not accompanied by
scintillation. It was shown that the average energy used to form
a carrier pair was equal to 19 eV and independent of the beta-particle energy. The scatter of the values of ε from sample to sample did not exceed 1 eV.

§ 16. Discussion of Experimental Data and a Theoretical Interpretation

The available data on the generation of nonequilibrium carrier
pairs by incident energetic charged particles are of great importance from the point of view of the detection and spectrometry
(energy determination) of particles. Unfortunately, the number
of published papers on ionization in semiconductors is small; for
such interesting substances as, for example, CdS, the value of
ε is not known sufficiently reliably. A theoretical interpretation
of the available data should, first of all, allow for the great similarity between the processes of impact ionization by photoelectrons and photoholes, when photons of energies much higher than
the forbidden bandwidth are absorbed, and of ionization by the
passage of a fast charged particle.

A theory developed recently by W. Shockley [20] made it possible to relate quantitatively these two phenomena in silicon and
germanium and to describe them (as well as the phenomenon of
impact ionization in strong fields) using a relatively simple model.
This theory was mentioned in Chapt. II in connection with an analysis of the data on the quantum yield rise in germanium and silicon
when the energies of the absorbed photons exceed about $3E_g$. According to the Shockley model, carriers with kinetic energies sufficient to produce secondary impact ionization should be scattered
mainly on phonons of the highest frequency. These phonons represent vibrations of a diamond-type lattice when the two face-
centered sublattices are displaced in opposite directions. The
wave vector for these vibrations is of zero length and is located

at the center of a Brillouin zone. They are known as the "Raman" vibrations of the crystal. The energy quantum, $\hbar \omega$, corresponding to their frequency, is denoted by E_R. The values of E_R were determined from the data on the scattering of cold neutrons by germanium and silicon crystals. These values were 0.063 eV for Si and 0.037 eV for Ge. According to Shockley, all the collisions of carriers with the lattice are accompanied by energy loss, which is a good approximation for silicon at room temperature when $kT < E_R/2$ and the high-frequency vibrations are practically not excited by thermal means. The value of E_R is one of the parameters in the Shockley model. The three selected constants are the following quantities:

a. E_i, which is the threshold energy of a carrier, taken from the bottom of a band; a carrier with energy $E > E_i$ is capable of forming a secondary electron-hole pair;

b. L_R is the mean free path of a carrier between collisions which are accompanied by the excitation of Raman vibrations;

c. $r = L_i/L_R$, where L_i is the mean free path, between impact ionization collisions, of a carrier whose energy is greater than E_i.

From the definition in c, it follows than an electron with an energy greater than E_i forms on the average r phonons in each act of ionization.

The probability that a carrier with energy E causes ionization before losing the excess energy $(E - E_i)$, and the ability to ionize, can be determined as follows. In order to lose the excess energy, the carrier must suffer

$$c = \frac{E - E_i}{E_R} \tag{3.7}$$

collisions accompanied by phonon emission. While doing this, the carrier travels a mean distance cL_R. The probability that it does not cause ionization over this distance is

$$\exp\left(-\frac{cL_R}{L_i}\right) = \exp\left(-\frac{E - E_i}{rE_R}\right). \tag{3.8}$$

The probability of one (or more) ionization acts during the slowing down of the carrier from the energy E to an energy less than E_i is, therefore,

$$P(E) = 1 - \exp\left(-\frac{E - E_i}{rE_R}\right). \qquad (3.9)$$

According to the data on impact ionization by photoelectrons and photoholes in Ge and Si (Chapt. II), the best agreement of the Shockley model with experiment is obtained by assuming that E_i = 1.1 eV for Si, and E_i = 0.68 eV for Ge. Using Bethe's theory of the ionization losses, Shockley assumed that the primary electrons and holes, generated when a high-energy charged particle is stopped in a semiconductor, have usually moderate energies, which are several times greater than E_i. After the primary ionization act, the energy possessed initially by electrons and holes is lost in the following way. Each act of impact ionization producing a pair absorbs energy E_i from the total kinetic energy of the system of nonequilibrium carriers. In addition, the energy rE_R is usually transformed into lattice vibrations. Finally, when a given carrier is no longer able to ionize, it still retains a residual kinetic energy E_F which is also transferred to lattice vibrations.

Obviously, the quantity E_F should be slightly smaller than E_i. The same quantity E_F represents the average kinetic energy of those nonequilibrium carriers whose kinetic energies on creation are smaller than E_i. Taking into account the excess energy E_F, the average energy ε per nonequilibrium carrier pair is given by

$$\varepsilon = 2E_F + E_i + rE_R. \qquad (3.10)$$

If we assume that carriers produced by impact ionization are equally likely to be located at any point of a Brillouin zone with an energy less than E_i, then for parabolic constant energy surfaces E_F = 0.6 E_i. Using the values of r determined by comparing the theoretical curves with the experimental spectral distribution of the photoionization quantum yield, Shockley obtained the following values for ε :

Parameter	Silicon	Germanium
E_i	1.1 eV	0.66 eV
rE_R	1.1 eV	2.1 eV
$\varepsilon = 2.2E_i + rE_R$	3.5 eV	3.6 eV

TABLE 5. Experimental Data on the Average Energy of Carrier-Pair Formation by Fast Particles and High-Energy Photons

Substance	Forbidden band-width, eV	Value of ε, eV	Radiation producing carrier pairs	Method of determining ε	Literature source
1. Germanium	0.72	3.0±0.4	Alpha-particles, E = 5 MeV	Measurement of current pulses due to ionization in the strong field of a p-n junction	[12]
		2.5	X rays; photo-electrons and Compton electrons on absorption of γ rays	Measurement of photocurrent between p- and n-type regions	[21] [22]
		3.7±0.4	Electrons, E = 5-15 keV	Measurement of multiplication coefficient, allowing for volume and surface recombination	[14] [20]
		4.5±1.5	Electrons, E = 400 keV		
2. Silicon	1.1	3.6	Alpha-particles, E = 5 MeV	Measurement of current pulses due to ionization in the strong field of a p-n junction	[13]
		4.2±0.6	Electrons, E = 1.5-30 keV	Measurement of multiplication coefficient, allowing for volume and surface recombination and for the "dead" layer at the surface	[15] [16]
3. Diamond	6	18-20	Beta particles	Conduction pulses in homogeneous crystals in a strong electric field	[18]
4. Cadmium sulfide	2.37	5-10	Alpha particles		[12]

To refine the theory further, one would have to find the value of E_F more rigorously, taking into account the band structure and the distribution of the residual energy $(E - E_i)$ between carriers after the generation of a pair.

Comparing the available experimental data for the ionization of germanium and silicon single crystals by charged particles, we can draw the following conclusions:

a. the energy ε, which represents the ratio of the total ionization losses to the number of generated nonequilibrium carrier pairs, is practically independent of the initial energy and the nature of the charged particle;

b. in the case of germanium and silicon, and (with less justification) in the case of diamond, there is a relationship between the value of ε and the forbidden bandwidth E_g: the ratio ε/E_g lies between 3 and 4; *

c. the available experimental data on ionization by the stopping of charged particles, on impact ionization by photoelectrons and holes, and on impact ionization in strong electric fields, are well accounted for by the Shockley theory.

Table 5 lists some data on the ionization processes in germanium, silicon, diamond (Kennedy's data), and CdS. Kennedy [18] interpreted his value $\varepsilon \approx 19$ eV completely differently from Shockley. Kennedy assumed that the energy ε consists of the forbidden bandwidth E_g, the average depth of a level in the valence band from which an electron is transferred to the conduction band (this depth is about 10 eV), and a small term which represents the excess kinetic energy of a free electron in the conduction band. This interpretation does not agree with the results for Ge and Si, since these crystals, like diamond, have valence bands of considerable width, and if the Kennedy interpretation were correct, the values of ε for Ge and Si would have been greater by a factor of 2-3 than the experimental values.

---

* Obviously, this applies also to CdS, for which $E_g \approx 2.4$ eV and $\varepsilon \approx 5$-10 eV.

Chapter IV

RADIATIVE RECOMBINATION IN SEMICONDUCTORS; POSSIBILITY OF THE AMPLIFICATION AND GENERATION OF LIGHT USING SEMICONDUCTORS

The direct recombination of an electron and a hole accompanied by photon emission, as well as radiative electron transitions to the local levels of defects or impurities, are frequently processes of the higher order. This applies to germanium and silicon where the overall recombination rate is governed by nonradiative transitions. Nevertheless, investigations of the spectra of recombination radiation, first detected by O. V. Losev in silicon carbide [1], have yielded important information on the band structure, on the energy levels of impurities and defects, and on the crystal lattice vibrations of germanium, silicon, and other semiconductors. Recently, it has been found that the relative probability of radiative recombination is close to unity in semiconductors with a narrow forbidden band, such as indium antimonide and lead sulfide.

Recombination radiation is one of the forms of luminescence; the investigation of this radiation in the case of "electrical" excitation by the injection of carriers through p-n junctions has been found to be a useful method in studying electroluminescence. Finally, the stimulated emission of semiconductor systems with "negative temperature" (population inversion)* may be regarded as a special case of radiative recombination of nonequilibrium carriers.

We shall consider in the next section radiative recombination in semiconductors, using the term "recombination radiation" because of its wide currency, although there is no essential difference between recombination radiation and luminescence.

One may expect that with the improvement of sensitive infrared receivers, the study of recombination radiation spectra

---

*See section 22, p. 132

will become the standard method of investigating impurities, band structure,and excited states of semiconducting crystals, along with optical absorption and photoconductivity measurements.

§17.   Theory of the Recombination of Electrons and Holes Accompanied by Photon Emission

The process of photon emission in the radiative recombination of an electron with a hole may be regarded as the converse of the absorption of a photon in the fundamental band, i.e., of the absorption of a photon which produces an electron-hole pair. Using this assumption, the data on the absorption of light in germanium in the wavelength range 1-2 μ, and the fact that the quantum yield is unity at λ = 1-2 μ [2], Van Roosbroeck and Shockley calculated the probability of the interband (direct) radiative recombination in semiconductors [3].

Under the conditions of thermodynamic equilibrium, the number of radiative recombination acts in the frequency range dν is equal to the number of acts of their generation by thermal radiation. The generation of pairs per unit time and volume is $P(\nu)\rho(\nu)d\nu$, where $\rho(\nu)d\nu$ is the density of photons in a crystal in the frequency range dν, and $P(\nu)$ is the probability of absorbing a photon of frequency ν in unit time. Integration over all the frequencies gives the total number of recombinations per unit volume in 1 sec, i.e., the recombination rate, R:

$$R = \int_\nu P(\nu)\rho(\nu)\,d\nu. \tag{4.1}$$

We shall show below that the main part of the above integral for any semiconductor is contributed by a relatively narrow range of frequencies near the fundamental absorption band edge.

For a steady-state deviation from thermal equilibrium, the radiative recombination rate may be given by the formula:

$$R_c = \frac{np}{n_i^2} R, \tag{4.2}$$

because the number of direct recombinations should be proportional to the product of the electron density n and the hole density p, and should be equal to R when np = n_i^2, which is true under

equilibrium conditions. From the expression for R_c, we can calculate the characteristic "decay time" (or lifetime) τ for relatively
small deviations from equilibrium on the assumption that recombination is only radiative. If Δn and Δp are small excess densities over the equilibrium values and if $\Delta n = \Delta p$, it follows from
Eq. (4.2) that

$$\left. \begin{array}{c} \Delta R_c = \left(\dfrac{\Delta n}{n} + \dfrac{\Delta p}{p} \right) R_c \\[2mm] \tau = \dfrac{np}{n+p} R_c^{-1}. \end{array} \right\} \qquad (4.3)$$

and

In a sufficiently heavily doped n- or p-type semiconductor the
equilibrium density of majority carriers is $n = n_0$ (or $p = p_0$), the
equilibrium density of minority carriers is low, and the radiative
recombination lifetimes are given by

$$\left. \begin{array}{l} \tau_p = \dfrac{p_0}{R} = 2\,\dfrac{p_0}{n_i}\,\tau_i = 2\,\dfrac{n_i}{n_0}\,\tau_i, \\[3mm] \tau_n = \dfrac{n_0}{R} = 2\,\dfrac{n_0}{n_i}\,\tau_i = 2\,\dfrac{n_i}{p_0}\,\tau_i, \end{array} \right\} \qquad (4.4)$$

where $\tau_i = n_i/2R$ is the lifetime in an intrinsic semiconductor.
The cross section σ representing the recombination interaction
between an electron and a hole is

$$\sigma = \frac{R}{n_0 p_0 v} = \frac{R}{n_i^2 v}, \qquad (4.5)$$

where v is the average thermal velocity of carriers.

To calculate the spectral density of photons $\rho(\nu)$ in the frequency range $d\nu$ we use Planck's formula in which the refractive
index dispersion must be allowed for in calculating the density
of electromagnetic field states:

$$2 \cdot 4\pi\, k^2\, dk = \frac{8\pi}{c^3} \left[n^{*3} \left(1 + \frac{d \ln n^*}{d \ln \nu} \right) \right] \nu^2\, d\nu, \qquad (4.6)$$

where n^* is the refractive index, c is the velocity of light, and
$k = n^* \nu / c$ is the wave number.

Hence,

$$\rho(\nu) = \frac{8\pi\nu^2}{c^3} \frac{n^{*3} \dfrac{d \ln n^*\nu}{d \ln \nu}}{\exp\left(\dfrac{h\nu}{kT}\right) - 1}. \tag{4.7}$$

The probability of the absorption of a photon, $P(\nu)$ can be written as

$$P(\nu) = \alpha \cdot v_g,$$

where $\alpha = 4\pi n^* \varkappa \nu/c$; here $\varkappa$ is the absorption index, which is determined experimentally from the spectral dependence of the transmission of the substance, † and the quantity v_g is the group velocity of the wave packet:

$$v_g = \frac{d\nu}{d\left(\dfrac{1}{\lambda}\right)} = \frac{c}{n^*} \frac{d \ln \nu}{d \ln n^* \nu}.$$

Thus,

$$P(\nu)\rho(\nu) = \alpha(\nu) v_g(\nu) \rho(\nu) = \frac{32\pi^2 \varkappa n^{*3}}{c^3} \frac{\nu^3}{\exp\left(\dfrac{h\nu}{kT}\right) - 1}. \tag{4.8}$$

The total radiative recombination rate R is, in accordance with Eq. (4.1), given by:

$$R = 32\pi^2 c \left(\frac{kT}{hc}\right)^4 \int_0^\infty \frac{n^{*3} \varkappa u^3 \, du}{e^u - 1}, \tag{4.9}$$

where $u = h\nu/kT$, or, in the form convenient for calculations:

$$R = 1.785 \cdot 10^{22} \left(\frac{T}{300}\right)^4 \int_0^\infty \frac{n^{*3} \varkappa u^3 \, du}{e^u - 1} \, \text{cm}^{-3} \cdot \text{sec}^{-1}. \tag{4.10}$$

The lower limit of integration is in practice not zero but u_0, corresponding to the "optical" width of the forbidden band, i.e., the

† It is assumed that the absorption of light is always accompanied by pair generation. For many semiconductors, reliable data on the absorption in the fundamental band are not yet available. In the case of crystals with defects, additional absorption or scattering of light may occur without pair generation.

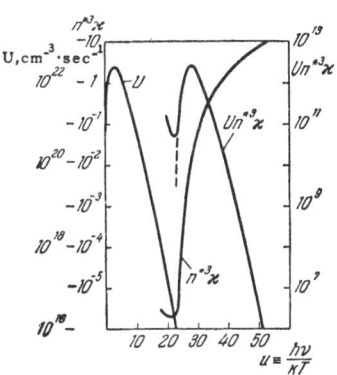

Fig. 51. Dependence, on $h\nu/kT$, of the quantities governing the probability of photon emission in the "fundamental" band of germanium.

minimum photoionization energy: $u_0 = E_{g0}/kT$. At longer wavelengths there are other mechanisms of light absorption; in particular, in semiconductors with a narrow forbidden band and at sufficiently high temperatures we may have intense absorption by carriers, which is not accompanied by pair generation.

To estimate quantitatively the radiative recombination rate in germanium, Van Roosbroeck and Shockley used the experimental data on the spectral dependence of $n*$ and $n*\varkappa$, published in [4]. Numerical integration of Eq. (4.10) gave for 300°K the value

$$R = 1.6 \cdot 10^{13} \text{ cm}^{-3} \text{sec}^{-1}.$$

Figure 51 shows the dependence of the quantity $P(\nu)\rho(\nu)$ on $u \equiv h\nu/kT$, as well as the dependences on u of the multipliers in the integral expression in Eq. (4.10): the quantity $U = 1.785 \times 10^{22} u^3/[\exp(u) - 1]$, which depends on the spectral density of photons, and the product $n*^3\varkappa$, which is governed by the properties of the semiconductor (germanium in our case). At 300°K, only a negligible fraction of photons has sufficient energy for the generation of pairs; the overlap of the curves representing U and $n*^3\varkappa$, in the region which lies approximately between $u = 25$ and $u = 32$, forms quite a sharp maximum. The rise of the $n*^3 \varkappa$ curve in the region $u < 22$ is related to the absorption of light by free carriers and therefore, in calculating the probability of radiative recombination and the shape of the radiation spectrum, this part of the curve was ignored and the resultant curve extrapolated downward (shown dashed). Using the value $n_i^2 = 3.1 \times 10^3 T^3 \exp(-9100/T)$, Van Roosbroeck and Shockley obtained $\tau_i = 0.75$ sec. Similar calculations based on more accurate experimental data on the absorption edge of germanium gave a value $\tau_i = 0.3$ sec.

It is known that even in the purest single crystals of intrinsic germanium the lifetime of nonequilibrium carriers does not ex-

TABLE 6. Radiative Recombination Lifetimes

E_{go}, eV	Substance	n_i, cm^{-3}	R, cm$^{-1} \cdot$sec^{-1}	τ_i, sec	τ obs, sec (max.)
1.1	Si	1.4×10^{10}	2×10^9	3.5	10^{-3}
0.7	Ge	2.4×10^{13}	3.7×10^{13}	0.3	10^{-3}
0.37	PbS	3×10^{15}	1.4×10^{20}	10^{-5}	9×10^{-5}
0.22	PbSe	2×10^{17}	3.3×10^{22}	3×10^{-6}	–
0.27	PbTe	6×10^{16}	1.8×10^{22}	1.7×10^{-6}	–
0.18	InSb	2.2×10^{22}	2.6×10^{22}	0.4×10^{-6}	0.1×10^{-6}

ceed several milliseconds. It has been shown that dislocations and structural defects act as recombination centers. Thus, even in crystals free of the usual "recombination" impurities – for example, copper – the lifetime is not governed by the radiative recombination.* This is even more pronounced in silicon for which the techniques of growing single crystals and of refining are far from perfect.

A theory based on the principle of detailed balancing allows us to draw the conclusion that in semiconductors with a narrow forbidden band, interband radiative recombination should be more likely than in germanium or silicon.

Burstein and Egli [5] used the published data on the absorption of light in several semiconductors with different forbidden bandwidths to estimate the radiative recombination rate at low excitation levels. The results of their calculations are given in Table 6.

The extreme right-hand column in Table 6 lists the longest experimentally observed lifetimes of electron-hole pairs. This table shows that in semiconductors with a narrow forbidden band, radiative recombination has a high probability; it is possible that in some cases this type of recombination determines the overall lifetime of minority carriers.

We have just considered the theory of recombination in which electrons are transferred from the conduction to the valence band. Using the same assumptions, we can consider the "impurity" radiative transitions, which are shown in Fig. 52. So far, no cal-

---

*Recently, Ge and Si single crystals free of dislocations have been prepared, so that we can expect longer lifetimes; the recombination of carriers in these crystals has not yet been studied.

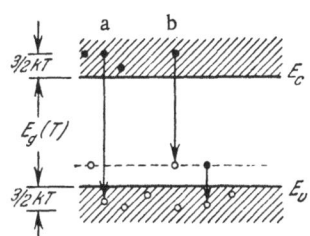

Fig. 52. Electron transitions in a semiconductor, accompanied by photon emission: (a) "intrinsic" transitions; (b) "impurity" transitions.

culations of the spectra of the "impurity" recombination radiation and of the probability of such radiation have been carried out for semiconductors such as Ge or Si, although the problem of the relationship between the radiative and nonradiative transitions has been widely discussed in the literature on luminescence [6] and on semiconductors [7]. Qualitative considerations suggest the existence of narrow impurity recombination radiation maxima; we shall show later that in many cases it is possible to observe this radiation.

§18. Experimental Methods of Exciting and Investigating Recombination Radiation Spectra

In work on recombination radiation spectra, the necessary excess carrier densities are established either by the excitation with short wavelengths [8] or by the injection of carriers by passing a current through a p-n junction in the forward direction [9, 10, 11]. The former method is particularly convenient in studying the intrinsic recombination radiation of germanium and silicon. A typical setup used in the study of the recombination radiation spectra of germanium is shown in Fig. 53. Light from an incandescent lamp with a tungsten filament passes through a water filter 10 cm thick and is focused on to a thin germanium plate. The radiation emitted on the opposite side of the plate is analyzed with a spectrometer. Haynes et al. [14] found that a water filter passed less than 10^{-10} of the radiant flux of wavelengths longer than 1.4μ, and that the germanium test plate transmitted less than 10^{-10} of the radiant flux of wavelengths shorter than 1.4μ, i.e., the light from the incandescent lamp did not reach the spectrometer and the recorded radiation could only be due to recombination or thermal processes. For silicon, one must use a water filter as well as a Zeiss KG-1 filter or a similar filter, because water transmits wavelengths up to 1.4μ and silicon is transparent in this region.

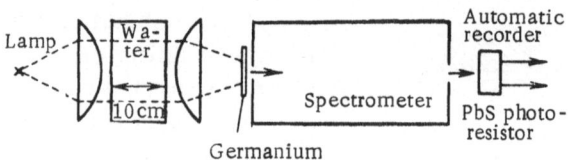

Fig. 53. Apparatus for the study of the recombination radiation spectra of germanium. Excitation is produced by lamps
generating wavelengths up to 1.4 μ.

An important problem is the emission of the radiation by the
sample. Under the simplest geometrical conditions (Fig. 54a),
a considerable part of the light flux suffers total internal reflection and does not reach the entry slit of the spectrometer. The
intensity of the radiation actually leaving the crystal can be raised
considerably by using samples in the form of a Weierstrass sphere
or "lens" [10]. In this case, the radiating volume is approximately
a hemisphere with a radius of the same order as the diffusion
length of minority carriers,with its center at the injection point
(Figs. 54b and 54c). In the case of injection through a small p-n
junction or through a point contact, a considerable fraction of carriers suffers recombination at the surface. By using the sharp
focusing of an optical system with a Weierstrass sphere, we can
investigate separately the "surface" and "volume" recombination
radiation. This sample geometry is convenient in studies of the
impurity radiation, because the host-crystal radiation, representing transitions between bands, is distorted by absorption.

Bolometers and thermocouples used as standard nonselective
receivers in infrared spectrometers are usually insufficiently
sensitive to record recombination radiation. Therefore, special
PbS photoresistors are frequently used for this purpose [12].

The noise level in the best of such photoresistors is very low,
and one can use their full sensitivity simply by cooling the radiation screens which protect them from the radiation background at
300°K. The modulation of the exciting light or of the injection
current can be used for synchronous detection of the signal, thus
raising the signal-to-noise ratio by increasing the time available
for recording the spectrum. P. Aigrain reported privately that
the infrared recombination radiation of germanium, due to an injection current of 1 A, was concentrated into a narrow beam by a

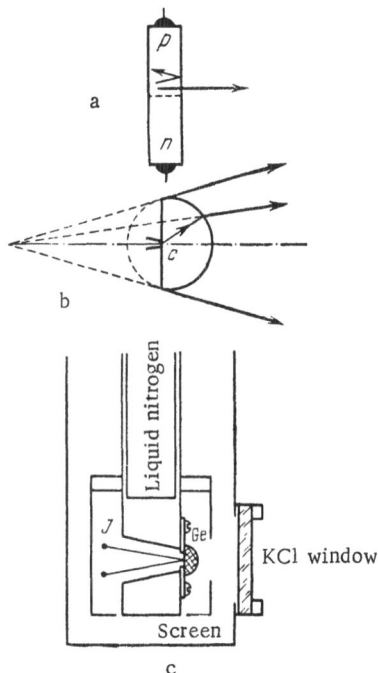

Fig. 54. a) Reduction of the recombination radiation intensity by total internal reflection. b) A germanium or silicon single crystal in the form of a Weierstrass sphere. The radiation is excited by injecting carriers through a contact. c) Apparatus used for studies on cooled crystals; J is the exciting current.

Weierstrass sphere and a mirror, and could be detected reliably at a distance of 1 km.

In investigating the recombination radiation of semiconductors with a forbidden band wider than 1.1 eV, e.g., silicon, one can use photographic plates or photomultipliers.

§ 19.   Recombination Radiation Spectra of Germanium

A.   "Intrinsic" Radiation

By analogy with the established concept of intrinsic conduction in semiconductors, due to electron transitions from the valence to conduction band, we shall call "intrinsic" the radiation which is produced by transitions in the opposite direction. The most detailed experimental data on intrinsic recombination radiation are available for germanium. Ignoring the earlier work on the detection of this radiation, we shall consider the results obtained by Haynes [8], who used the photoexcitation method and thin germanium plates. Even in the case of very thin germanium plates (Haynes used plates from 1.2×10^{-2} to 1.3×10^{-3} cm thick), the dependence of the radiation intensity on the wavelength in the intrinsic radiation region had to be corrected for absorption. When the carrier diffusion length was considerably greater than the sample thickness and the surface recombination velocity was quite small (this was achieved by special etching), Haynes assumed that the volume lifetime and carrier density were the same throughout the sample. The initial recombination radiation flux $I_{0\lambda}$ of wavelength λ

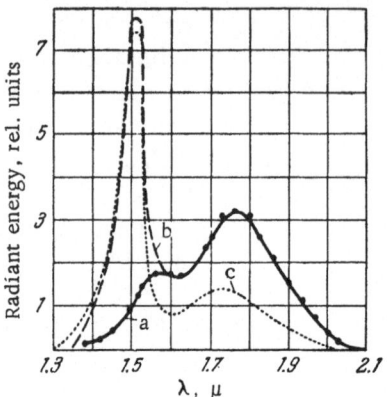

Fig. 55. Spectrum of the intrinsic recombination radiation of a thin germanium
plate. a) Experimental curve; b) experimental curve with allowance for self-absorption; c) curve plotted using Dash's absorption data [13] and Van Roosbroeck and
Shockley's theory.

and the flux detected just outside the plate, I_λ, were related by

$$I_{0\lambda} = \frac{I_\lambda \alpha_\lambda d}{1 - e^{-\alpha_\lambda d}},$$

where α_λ is the absorption coefficient which depends on the wavelength λ, d is the sample thickness; $\alpha_\lambda d > 1$. The above formula is approximate since it does not allow for reflection at the
germanium-air boundary; however, the variation of the reflection coefficient in the 1.4–2.1 μ region is small compared with
the variation of the absorption coefficient.

After analyzing the experimental data and correcting the curve,
a sharp maximum was found at $\lambda = 1.52 \mu$ (0.81 eV), in addition
to the maximum at 1.75 μ (0.70 eV), which was found in earlier
work and which matches the forbidden bandwidth of germanium at
300°K (Fig. 55). Initially, there had been some difficulties in the
interpretation of the 1.52 μ maximum; recent experimental data
on the existence of a prominent structure in the absorption band
edge of germanium (cf. Chapt. I) showed that the use of the earlier results of Briggs (obtained for thin deposited films) in discussing germanium single crystals was not justified. Carrying
out a numerical integration of Eq. (4.10) of the Van Roosbroeck–

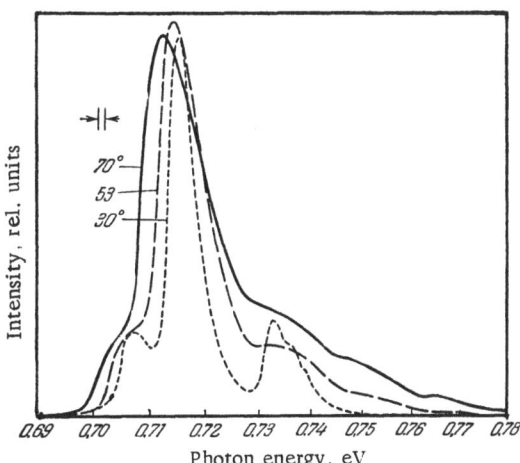

Fig. 56. Structure of the intrinsic recombination radiation spectrum of germanium, associated with indirect
transitions.

Shockley theory, Haynes obtained good agreement between the
theory and his own results. The continuous curve with experimental points, denoted by a in Fig. 55, represents the experimental
dependence of the radiant energy on the wavelength for a very thin
plate (d = 1.3 × 10⁻³ cm). The curve denoted by b, plotted with
an allowance for absorption, left no doubt about the existence of
the second maximum, which was barely noticeable in curve a.
Bearing in mind the fact that multiple internal reflections raise
the relative intensity of the radiation of longer wavelengths, the
agreement between the experimental and theoretical curves must
be regarded as satisfactory. The maximum at 1.52 μ was obviously due to vertical transitions at k = 0, as suggested in studies
of the absorption processes [13], and the maximum at 1.75 μ
(0.70 eV) was due to transitions with phonon participation. Thus,
the (000) minimum of germanium lay only 0.1 eV above the (111)
minimum.

On cooling the germanium, its radiation bands became narrower and shifted toward short wavelengths. The direction and
magnitude of the shift agreed with the change in the forbidden bandwidth observed by Fan and other workers.

Recently [14], fine structure has been observed in the intrinsic
recombination spectra of germanium and silicon, where it has been

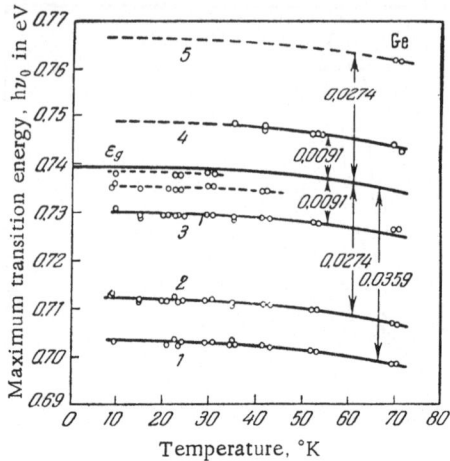

Fig. 57. Variation of the energy $h\nu_0$ with temperature. 1) Emission of an optical transverse
phonon; 2) emission of an acoustic longitudinal
phonon; 3) emission of an acoustic transverse
phonon; 4) absorption of an acoustic transverse
phonon; 5) absorption of an acoustic longitudinal
phonon.

found that the "principal" maximum at long wavelengths is accompanied by auxiliary maxima. According to Haynes, Lax, and Flood
[14], the various maxima represent the energies of various types
of phonon, i.e., the energies of the interaction of an electron
with the lattice in a radiative transition. The structure is observed below 30°K, and is particularly clear at the temperature
of liquid helium.

The intrinsic recombination radiation spectrum of germanium
in the indirect transition region is shown in Fig. 56 for temperatures from 70 to 30°K. At 30°K, this spectrum has three clear
lines due to the emission of phonons of various energies. The energies $h\nu_0$, obtained by extrapolating to zero intensity the falling
parts of the curve on the short-wavelength side of the minima, are:
0.730 eV, which represents the emission of a transverse acoustic
phonon; 0.712 eV, which represents the emission of a longitudinal
acoustic phonon; and 0.703 eV, which corresponds to a transverse
optical phonon. At higher temperatures (53°K), we can distinguish
bands representing both the emission and absorption of transverse

TABLE 7. Phonon Energies, in eV, Corresponding to Wave
Vector k_0 [14].

Element	Transverse acoustic phonon	Longitudinal acoustic phonon	Transverse optical phonon	Longitudinal optical phonon
Germanium	0.0091	0.0274	–	0.0359
Silicon	0.016	0.055	0.083	0.119

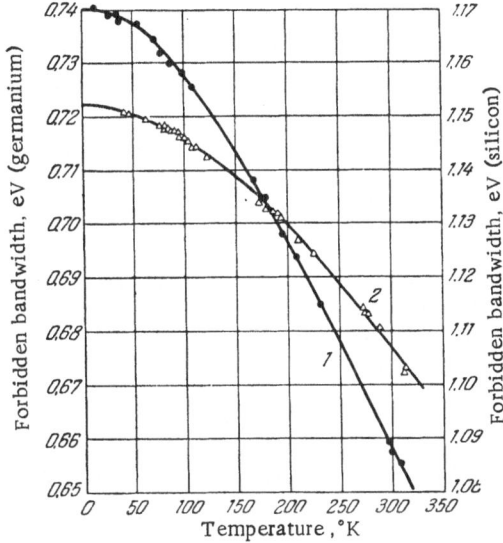

Fig. 58. Temperature dependence of the optical forbidden bandwidth of germanium (1) and silicon (2),
deduced from their recombination radiation spectra [14].

acoustic phonons. The energy $h\nu_0$, corresponding to a transition
accompanied by the emission of a transverse acoustic phonon, is
0.728 eV, while for a transition with the absorption of the same
type of phonon, $h\nu_0$ is 0.746 eV.

At 70°K, it is possible to observe a line corresponding to the
absorption of a longitudinal acoustic phonon with an energy $h\nu_0$
= 0.726 eV. Lines representing transitions with the participation
of longitudinal optical phonons were not detected by Haynes, Lax,
and Flood [14]; this was perhaps due to the fact that these lines

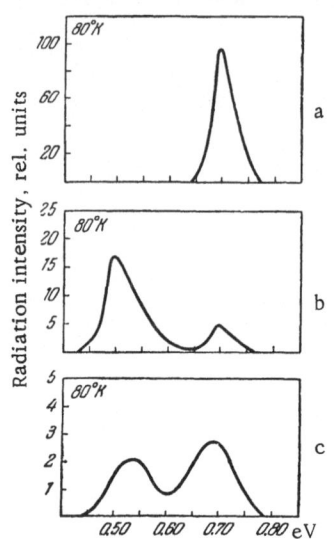

Fig. 59. Recombination radiation spectra of plastically deformed germanium. a) Crystal not subjected to deformation; b) crystal twisted through 1°; c) crystal twisted through 1° and annealed at 800°C (measurements were carried out at 80°K).

should lie between intense lines representing transitions with the participation of acoustic and transverse optical phonons. The variation of $h\nu_0$ with temperature is shown in Fig. 57. Above 70°K, it is possible to determine the value of $h\nu_0$ only for the transitions accompanied by the emission or absorption of a longitudinal acoustic phonon. If these transitions represent the recombination of electrons and holes, the energy E_g corresponds to the middle point between the values of $h\nu_0$ representing the emission and absorption of a longitudinal acoustic phonon, as shown in Fig. 57. The lines close to E_g, whose positions are represented by the dashed curves, may be due to the recombination of excitons in the ground and first excited states, without phonon participation. This may occur when excitons are scattered by neutral impurity atoms.

Table 7 lists the values of the phonon energies for germanium and silicon, deduced by Haynes et al. [14] from an analysis of the fine structure of recombination radiation spectra.

The results of Haynes et al. [14] are in good agreement with those obtained by Macfarlane et al. [15] (who investigated the optical absorption) and with the data on cold-neutron scattering in germanium crystals [16].

The data on the forbidden bandwidth, obtained for Ge and Si by analyzing the fine structure of their recombination radiation spectra at various temperatures, are given in Fig. 58.

If electrons and holes form excitons before recombination, the forbidden bandwidth is greater than that given in Fig. 58 by an amount representing the binding energy of excitons.

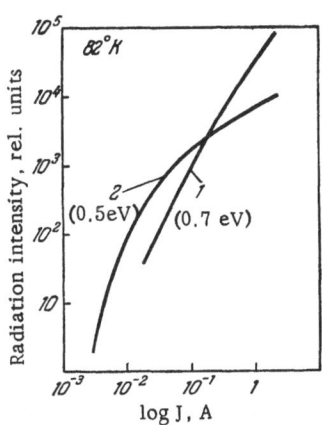

Fig. 60. Dependence of the intensity of the intrinsic (1) and impurity (2) recombination radiation
of germanium on the excitation
intensity [17].

B. Impurity Radiation of Germanium

Impurity recombination radiation, corresponding to electron
transitions of type b in Fig. 52,
may result from recombination
through levels lying in the forbidden
band. It is relatively easy to detect this radiation because of the
high transparency of semiconductors outside their fundamental absorption band; on the other hand,
it is difficult to find sufficiently
sensitive receivers for wavelengths
longer than 6-8 μ. The radiation
corresponding to electron transitions from un-ionized group V donors
in germanium should have a maxi-
mum about 0.01 eV from the "interband" maximum, therefore it
has not yet been distinguished from the fundamental band. However, the first experiments of French physicists using Weierstrass spheres indicated the presence of a clear maximum at
2.45 μ at 77°K [10]. Aigrain et al. associated this maximum
with recombination centers in germanium which have a capture
level 0.22 eV from one of the bands. The determination of the
positions of the recombination center levels from the temperature
dependence of the carrier lifetime meets with serious difficulties.
Therefore, impurity recombination radiation spectra may yield
very valuable independent data. Investigating the recombination
radiation of plastically deformed germanium, Newman [17] showed
that the maximum close to 2.4 μ corresponds to energy positions
of dislocations. His initial purpose was to detect radiative transitions associated with deep levels of several impurities (Au, Mn,
Ni, Fe, Cu, etc.) in germanium. He prepared crystals with p-n
junctions which were then doped with these impurities. It is known
that the solubility of these elements in germanium is low. Irrespective of the nature of the impurity, the maximum at 2.4 μ
was observed at 80°K in all the crystals whose p-n junctions were
prepared by the fusion of indium; this maximum was absent in
crystals whose junctions were established during the growth of

single crystals. Assuming that the 2.4 μ maximum was due to defects or dislocations generated in the recrystallization zone by the fusion of indium, and not due to impurities, Newman investigated the recombination radiation of crystals plastically deformed by torsional sheer at 550°C under conditions which eliminated the possibility of contamination with accidental impurities. Annealing at 800°C for 15 hr (which, according to Tweet [18], should have been sufficient for the elimination of thermal acceptors) displaced somewhat the maxima (from 0.50 to 0.53 eV) and changed the relative intensity of the intrinsic and impurity radiation maxima (Fig. 59). The introduction of copper (about 10^{15} cm^{-3}) produced a new maximum, corresponding to 0.59 eV, which, as pointed out by Newman himself, did not match the position of any of the three known acceptor levels of Cu in Ge (0.26 eV from the conduction band; 0.04 eV and 0.33 eV from the valence band). The 0.59 eV maximum could have been the result of the superposition of the radiation spectrum of copper and that associated with transitions to dislocation levels.

In conclusion, we shall quote some data on the relationship between the radiation intensity and the excitation. It follows from the theory that the velocity of direct radiative recombination at small departures from equilibrium should be proportional to the product np. The excess carrier density is in most cases determined indirectly; the quantity used to represent the deviation from thermal equilibrium is often the forward current through a p-n junction. This approach represents a rough approximation, because, particularly at high forward current densities J, the carrier lifetime and "injection coefficient" depend on the current density and the proportionality of $\Delta n \propto J$ is disturbed. Nevertheless, the dependence of the "intrinsic" radiation intensity on the current (curve 1 in Fig. 60) was nearly quadratic in Newman's work [17]. The same dependence was obtained by other investigators [20]. It is more difficult to interpret the results of Newman obtained for the impurity ("dislocation") radiation. The impurity radiation curve (2), obtained partly with a monochromator and partly with a thick germanium filter, which completely absorbed the "intrinsic" radiation, shows that at very low injection current the intensity was proportional to the fifth power of the current. Newman ascribed this anomaly to the presence, near the junction, of a very thin layer in which the carrier lifetime is very short.

§ 20.   Radiative Recombination in Silicon

In 1952 Haynes showed [9] that the recombination of electrons and holes in silicon produced radiation with a maximum near 1.1μ. After a considerable improvement of the technique of measuring recombination radiation spectra (using optical excitation at short wavelengths and uniform silicon samples), Haynes detected a fine structure which was in many respects similar to the analogous structure in the spectra of germanium. The temperature dependence of the forbidden bandwidth E_g of silicon is shown in Fig. 58; the phonon energies are listed in Table 7 (Section 19).

The intrinsic and impurity radiation were distinguished reliably in the spectra of silicon [21]. In studies of the impurity radiation, p-n junctions were prepared by doping during the growth of single crystals. The radiation was analyzed with a Perkin-Elmer spectrometer; a PbS photoresistor was used as the receiver. A correction for the dependence on wavelength of the photoresistor sensitivity and spectrometer transmission was made by comparison with the spectrum of a tungsten lamp operating at a known temperature. The recombination radiation spectrum of silicon containing boron and arsenic impurities is shown in Fig. 61. At room temperature the maximum of the radiation spectrum, corrected for self-absorption, corresponds to 1.088 eV. The continuous curve shows the radiation spectrum at 77°K. This spectrum consists of a very narrow band with a maximum at 1.10 eV and two weaker bands with maxima at 1.072 and 1.038 eV. Both the radiation observed at room temperature and the band with a maximum at 1.1 eV (observed at 77°K) reflect the intrinsic properties of silicon; it has been found that they are independent of the nature of impurities and are in agreement with the calculations based on the optical constants of silicon and on the detailed balancing principle. The position of the radiation band corresponds to the first sharp rise in the fundamental absorption band of silicon [13]. This region of the radiation band is due to nonvertical (indirect) electron transitions from the (1, 0, 0) minimum to the upper part of the valence band with the participation of phonons. Consequently, this radiation may be called "indirect intrinsic." No radiation due to vertical (direct) transitions has been found for silicon, which differs basically in this respect from germanium.

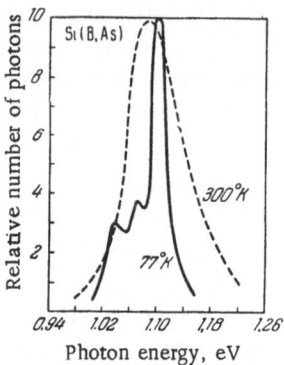

Fig. 61. Recombination radiation spectrum of silicon containing boron and arsenic impurities at room temperature and at liquid nitrogen temperature.

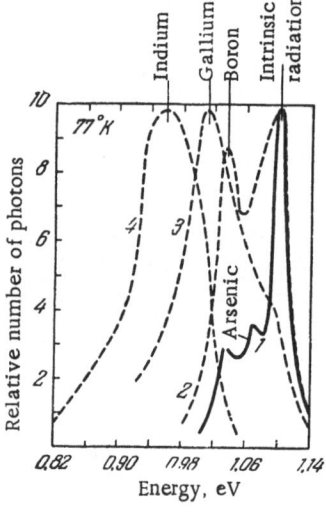

Fig. 62. Effect of the nature and concentration of impurities in silicon on its recombination radiation spectra.

The additional maxima at 1.072 and 1.038 eV are related uniquely to the type of impurity in silicon. Figure 52 shows intrinsic (a) and impurity (b) radiative transitions; the holes, denoted by open circles, lie – on the average – $(3/2)kT$ below the top of the valence band. The acceptor levels, shown as a dashed line, are partly ionized. In this case, the injected electrons (black dots in the conduction band) may recombine only with the holes in the valence band or with the holes at the acceptor levels. In the former case (intrinsic radiation), the photon has an energy $h\nu_1 = E_g(T) + 3kT \underset{(+)}{-} \hbar\omega$, where $\hbar\omega$ is the energy of a phonon which is needed to satisfy the law of conservation of momentum. The transitions should be usually accompanied by the emission of phonons, since the number of phonons which can be absorbed in such a process is very small.

In the impurity radiation case the participation of phonons is also essential, at least for the impurity levels with low ionization energies. The emitted photon should have the energy $h\nu_2 = E_g(T) - E_i + (3/2)kT \underset{(+)}{-} \hbar\omega$, where E_i is the ionization energy of the impurity center.

The influence of impurities in silicon on the recombination radiation is shown in Fig. 62. The continuous curve is identical with the curve for 77°K, given in Fig. 61.

Curve 2 represents an increase of the concentration of boron by a factor of 50, the concentration of arsenic being kept fixed. The steep rise of the maximum at 1.039 eV

TABLE 8

Element	E_i (from recomb. rad.), eV	E_i (thermal), eV	E_i (optical), eV
B	0.051	0.045	0.046
Ga	0.075	0.065	0.071
In	0.13	0.16	0.16
As	0.018	0.049	0.056

indicates that the radiation is due to the recombination of electrons and holes of un–ionized boron atoms. The nature of the maximum at 1.072 eV is less clear; it is suggested that this maximum is associated with the recombination of holes with electrons at un-ionized arsenic atoms. The substitution of gallium for boron dis-places the maximum to 1.015 eV (curve 3). The greatest displace-ment of the maximum – to 0.960 eV – is obtained when indium is the main impurity (curve 4). According to the above interpreta-tion of this displacement, $E_i = h\nu_1 - h\nu_2 - (3/2)kT$.

Table 8 lists the following quantities: the ionization energies of impurities in silicon, deduced from the recombination radiation spectra; the activation energies, deduced in [22] from the temper-ature dependence of the electrical conductivity and Hall effect; the ionization energies, found from the photoconductivity and absorp-tion spectra [23].

The position of the maximum ascribed to the donor levels of arsenic exhibited a large scatter whose origin was not clear. Com-paring the fine structure of the recombination radiation spectra of almost intrinsic silicon and of silicon containing impurities of groups III and V, Haynes observed below 25°K very narrow lines whose intensity was proportional to the impurity concentration. Narrowing the spectrograph slit, Haynes established that the widths of these lines did not exceed 0.0005 eV. According to him, these lines were associated with the radiative recombination of almost stationary electrons and holes since any thermal excitation energy of the carriers would have broadened the lines in accordance with the Boltzmann distribution. The relationship between the radia-tion intensity and the impurity concentration suggested that elec-trons and holes formed, before recombination, complexes with impurity atoms (for example, with arsenic atoms). The energy difference between the two narrow lines was equal to the energy of a transverse optical phonon. Thus, the recombination may have occurred either without phonon participation or with phonon emission. In the former case, the law of conservation of momentum was satisfied by momentum transfer to the whole crystal via impurity atoms, representing low energy loss.

§ 21.   Radiative Recombination in Semiconducting Compounds (InSb, GaSb, InP, PbS)

Moss and Hawkins [24, 25] determined experimentally the total number of recombination radiation photons for indium antimonide. As in the case of germanium and silicon, the position of the recombination radiation band of InSb corresponded to the forbidden bandwidth ($E_g \approx 0.18$ eV).

The radiation sources were thin (about 12 μ) plates of single-crystal indium antimonide refined by zone melting. They were excited with light; the glass prisms, used to focus the light of a tungsten filament lamp on the test plates, cut off completely the long wavelengths. The InSb radiation was focused onto the slit of a monochromator by a mirror of aperture 1 : 0.8. A sensitive thermopile was used as the radiation receiver. The recombination radiation spectrum of InSb is shown in Fig. 63. Allowing for the absorption and reflection losses, Moss and Hawkins concluded that radiative recombinations accounted for 20% of all recombinations but the probable error was ±50%. This result was in agreement with the theoretical predictions (Table 6). The number of photons emitted into the surrounding medium in 1 sec amounted to 10^{14} in Moss's experiments; the greatest losses, as in the case of tests on Ge and Si plates, were due to the total internal reflection and absorption of radiation.

One of the possible methods of exciting recombination radiation is by the impact multiplication of carriers in an avalanche breakdown. Because of its narrow forbidden band, indium antimonide breaks down in fields over 200 V/cm.

Basov, Osipov, and Khvoshchev [26] observed the recombination radiation of InSb on the application of electric-field pulses to a sample at 78°K. The pulse duration was about 3 μsec and the repetition frequency up to 50 cps. The radiation vanished on heating the sample to 120-180°K. The rise and decay times of the radiation pulse did not exceed 1 μsec, indicating that the radiation

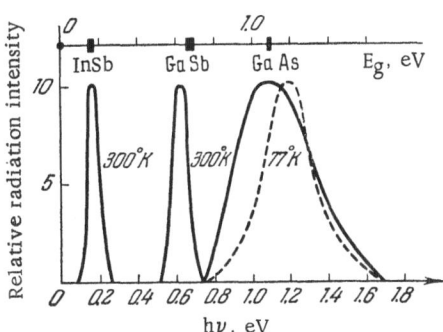

Fig. 63. Recombination radiation spectra of intermetallic compounds.

was not due to heating of the crystal. The radiation spectrum
had a maximum at 5.3 μ and the half-width of the maximum was
0.25 μ.

Braunstein [27] investigated the recombination radiation spectra of the intermetallic compounds GaSb, GaAs, and InP, and
found that the radiation band in each case coincided approximately
with the absorption band edge (Fig. 63). To excite the radiation,
Braunstein injected minority carriers from rectifying point contacts or from large-area contacts made by depositing a silver
paste on the etched surface of a crystal. In comparing the positions of the radiation band maxima with the values of E_g of these
intermetallic compounds, one must bear in mind that, because of
insufficient purification of the crystals and their lack of perfection it is difficult to determine the "optical" forbidden bandwidth
because of the presence of other types of absorption not associated
with carrier pair generation (Chapt. I). In such cases, additional
data are necessary to determine whether the observed radiation
bands are of the impurity or intrinsic type. For example, the
radiation maximum of GaSb at 300°K corresponds to 0.625 eV,
while the forbidden bandwidth is 0.67 eV [28]. There is also a
similar difference for GaAs: the radiation maximum of GaAs corresponds to 1.10 eV at 300°K; the forbidden bandwidth of the same
substance is 1.35 eV [29]. The intensity of the radiation emitted
by the three compounds studied by Braunstein was proportional to
the forward current across the rectifying contact, which was due
to the high majority carrier density (about 10^{17} cm^{-3}).

Recently, Nasledov, Rogachev, Ryvkin, and Tsarenkov [41] obtained more detailed information on the intrinsic and impurity recombination radiation spectra of gallium arsenide. According to their results, a narrow intrinsic radiation maximum was observed near 1.47 eV at 77°K; photons of lower energy were due to transitions through impurity centers.

The recombination radiation of lead sulfide was observed by Galkin and Korolev [30], who investigated polycrystalline layers of this compound. In a latter study, Scanlon [31] obtained data on the absorption band edge of PbS single crystals and calculated the carrier lifetime in the case of radiative recombination: this lifetime was about 6×10^{-5} sec at 300°K. Obviously, this result is more accurate than the value quoted in Table 6 (about 10^{-5} sec). The data on the band structure of PbS do not exclude the possibility of two intrinsic radiation maxima, corresponding to direct and indirect electron transitions.

§ 22. Feasibility of Producing Population Inversion (Negative Temperatures) in Semiconductors. Amplification and Generation of Coherent Radiation

It has just been shown that, in general, any semiconductor containing nonequilibrium electrons and holes is a source of more or less monochromatic recombination radiation, i.e., it is a converter of "nonthermal" excitation energy into light. However, under normal conditions, this radiation is of very low intensity and incoherent.

The rapid development of a new branch of science, known as quantum radiophysics, has established that, in principle, it is possible to produce powerful monochromatic sources emitting either radiowaves or radiation in the visible or infrared regions [32, 33].

The operation of these new radiation sources (quantum generators or masers) is based on the principle of stimulated emission of electromagnetic waves by excited quantum systems. In contrast to the sources available before, the stimulated radiation principle makes it possible to cohere the radiation of an assembly of quantum microsystems by establishing a special thermodynamic nonequilibrium state known as the population inversion or negative temperature state.

One of the main problems in the design of masers is the selection of the substance in which population inversion can be established. In the infrared and visible regions intensive studies
are proceeding on the use of gases, luminescent crystals and semiconductors.

Theoretical and experimental investigations of masers have
attracted the attention of a great many Soviet and foreign workers.
It seems appropriate to include the present short section on masers
in the chapter dealing with the recombination radiation of semiconductors. If a semiconductor is used as the working substance
of a maser, all the processes in the crystal, beginning with the
excitation of nonequilibrium carriers and ending with the emission
of light, may be considered as special cases of radiative recombination.

A. Population Inversion Concept

The term "population inversion" is used to describe a situa-
tion when the population of the upper of two energy levels is higher
than the population of the lower. If a system (a substance) in the
state of population inversion is subjected to external radiation of
frequency equal to the frequency of the transition between the two
energy levels, the number of radiative transitions stimulated by
the external radiation is greater than the number of absorption
transitions. Thus, such a system can amplify the external radiation [32]. We shall consider an assembly of identical particles
(molecules or atoms). At thermodynamic equilibrium, the function which represents the distribution of the particles over the energy levels, f_i, is given by the statistical physics formulas:

the Bose-Einstein formula:

$$f_i = \frac{1}{e^{\frac{E_i - \mu}{kT}} - 1}, \tag{4.11}$$

or the Fermi-Dirac formula:

$$f_i = \frac{1}{e^{\frac{E_i - \mu}{kT}} + 1}. \tag{4.12}$$

The distribution function f_i gives the number of particles in a state i of energy E_i; μ is the chemical potential. In the limiting case, when $f_i \ll 1$, these formulas reduce to the classical Boltzmann distribution

$$f_i = e^{\frac{\mu - E_i}{kT}} .$$ (4.13)

In all cases, the probability of occupation of any given energy level decreases with increase of the energy of this level, i.e., if E_i > E_k, then $f_i < f_k$. This property is characteristic of all the distribution functions in systems at thermodynamic equilibrium.

The concept of temperature is, strictly speaking, inapplicable to systems which are not at thermodynamic equilibrium. However, it is convenient to introduce a quantity T_{ik} when discussing the interaction of such systems with electromagnetic radiation. This quantity is defined by

$$\frac{f_i}{f_k} = e^{-\frac{E_i - E_k}{kT_{ik}}} .$$ (4.14)

It may be called the "effective temperature" for levels i and k. The introduction of the effective temperature concept allows us to divide nonequilibrium systems into two classes:

a. systems for which the following inequalities are satisfied for any two energy levels i and k selected from the whole range of possible levels:

$$f_i < f_k; \quad E_i > E_k;$$ (4.15)

b. systems for which the following inequalities are satisfied for at least one pair of levels:

$$f_i > f_k; \quad E_i > E_k.$$ (4.16)

For systems of class (a), the effective temperature T_{ik} is positive. For systems of class (b), this quantity is negative, i.e., there is a population inversion.

When electromagnetic radiation interacts with matter, the atoms of the latter may absorb radiation quanta and thus be transferred to states of higher energy or they may emit quanta and so go over into lower energy states. The emission consists of: (1)

spontaneous transitions (which are analogous to the similar phenomena of spontaneous decay of unstable nuclei); and (2) stimulated
(induced) transitions to the ground state due to the action of an
external electromagnetic field on an atom. In the latter case, the
emitted quanta are identical with the quanta which are causing
stimulated emission. The spontaneous transitions are not re-
levant to the process of amplification of electromagnetic waves
but they govern noise in quantum systems.

Let us assume that particles, obeying Bose or Fermi statistics and having two energy levels E_i and E_k ($E_i > E_k$), interact
with radiation of frequency $\omega = (E_i - E_k)/\hbar$.

The number of transitions per unit time caused by the absorption of quanta of the external radiation is

$$I^- = n w_{ki} f_k (1 - f_i), \tag{4.17}$$

where w_{ki} is the probability of a transition from state k to state i;
and n is the number of photons.

The number of transitions accompanied by the emission of
quanta is

$$I^+ = (n + 1) w_{ik} f_i (1 - f_k), \tag{4.18}$$

where w_{ik} is the probability of a transition from state i to state k.
The multiplier $(n + 1)$ in the latter expression allows for stimulated
and spontaneous transitions. According to the principle of detailed
balancing, the probabilities of the two processes are equal, i.e.,
$w_{ik} = w_{ki} = w$, and the total number of emission transitions may
be expressed as follows:

$$I^+ - I^- = wn (f_i - f_k) + w f_i (1 - f_k). \tag{4.19}$$

It follows from the above formula that if atoms are at a temperature $T_{ik} > 0$, the total number of transitions is zero only when
the temperatures of radiation and matter are equal. If the quantity I^+ is greater than the quantity I^- the system emits spontaneous radiation which cannot be used for the generation and amplification of electromagnetic radiation.

A completely new result is obtained if we assume that the
substance is in a state of population inversion. Then, the first
term of Eq. (4.19), representing the stimulated (coherent) emis-

sion of quanta, is positive and the system can amplify electromagnetic radiation passing through it.

If several energy levels are present in a system, so that radiation of a given frequency can cause transitions between pairs
of different levels, the condition for the amplification of radiation* becomes:

$$\sum_{ik} w_{ik} \, (f_i - f_k) > 0, \qquad\qquad (4.20)$$

where the summation is carried out over those states i and k,
which satisfy the condition $E_i - E_k = \hbar\omega$.

Obviously, if population inversion (negative temperature) can
be obtained with any pair of levels E_i and E_k, then the condition
(4.20) is satisfied automatically. However, even when the temperature is positive for some energy level pairs, the inequality
(4.20) may still be satisfied. This point is particularly important
in the case of semiconductors when the absorption of quanta by free
carriers may complicate considerably the amplification or generation of radiation.

B. Feasibility of Producing Population Inversion in Semiconductors

Several methods have recently been proposed for achieving
population inversion in semiconductors [32, 34, 35]. The interest in the possibility of using semiconductors as the working substance in masers is due to the fact that in semiconductors we can,
in principle, have the highest density of active particles, as well
as a high degree of perfection and purity of the crystals – such as
that obtainable for germanium, silicon, and indium antimonide.
The fact that the physical processes in these semiconductors have
been studied very intensively is an added advantage.

Basov, Vul, and Popov [34] suggested that population inversion may be obtained by impact ionization, which occurs on the
application of a strong electric field to a semiconductor (above a
certain critical value of the field the number of mobile carriers
in the bands increases sharply). Processes of this kind may be
related both to impact generation of pairs (knocking out valence

---

* Sometimes the term "negative absorption" is used in place of amplification; this
follows from the analogy between the formulas describing amplification and absorption, differing only in the sign of the index of the exponential function.

electrons) and to impact ionization of impurity atoms. Another possible mechanism which increases carrier density is the tunnel effect.

However, the use of a strong electric field, by means of which the carrier density can be sharply increased, does not give rise to large populations of electrons in the conduction band and holes in the valence band [36]. This is because the kinetic energy of carriers increases strongly in a field ("heating" of carriers). Consequently the number of energy levels which contain carriers increases to approximately the same extent. Therefore, one would expect that population inversion would be produced only after rapid removal of the electric field. Then, on the one hand, carriers are slowed down to energies corresponding to the lattice temperature, and, on the other hand, electrons recombine with holes. If the time taken by carriers to slow down is considerably shorter than their lifetime in the bands, then, until electrons and holes recombine, the population of levels near the bottom of the conduction band and the top of the valence band becomes very high and, at certain values of the nonequilibrium carrier density, population inversion may be achieved [35].

Among other methods suggested for producing population inversion, one may mention the use of electron transitions between Landau levels in strong magnetic fields when these levels are no longer equidistant [37]. This method, and other suggestions which have not yet been realized in practice, are critically reviewed in [35] and in Troup's monograph "Masers, Microwave Amplification and Oscillation by Stimulated Emission" [33].

At the beginning of 1961, Basov, Krokhin, and Popov* suggested the injection of minority carriers across a p-n junction at the boundary of a degenerate semiconductor for producing population inversion of carriers in a thin layer near the junction.

In 1962, R. N. Hall and his group at General Electric published the results of experiments in which coherent recombination radiation was observed on the injection of carriers by a forward current in GaAs crystals with p-n junctions [42]. The observations were quite conclusive because of the orientation of the radiation, the presence of a "threshold" value of the current density at which the intensity of emission rose sharply, and the con-

* N. G. Basov, O. N. Krokhin, and Yu. M. Popov, ZhETF 40:1879 (1961).

siderable (by a factor of 10) reduction of the width of the emission band. Inverse population was obtained by injecting carriers from "degenerate" n- and p-type regions into the junction region. The relatively high probability of radiative recombination by interband transitions and the fact that the photon energy of the emitted radiation was less than the threshold, representing the fundamental band edge in the degenerate regions (Chapt. I), made it easier to detect the effect, which could be used as the basis of operation of a semiconducting maser. Similar results were obtained at the P. N. Lebedev Physics Institute of the USSR Academy of Sciences [43].

In conclusion, we shall consider the recently discussed [35, 38, 39] possibility of achieving population inversion in semiconductors – such as germanium and silicon – by utilizing the characteristic features of nonvertical (indirect) electron transitions, in which the laws of conservation of momentum and energy are satisfied by the participation of lattice vibrations, i.e., by the emission or absorption of phonons (Chapt. I).

We shall consider the recombination of an electron and a hole accompanied by the simultaneous emission of a photon and a phonon. Obviously, if the electron and the hole have low kinetic energies, the resultant recombination radiation corresponds to the long-wavelength edge of the absorption band of the semiconductor. The inverse transition represents the generation of an electron and a hole with the simultaneous absorption of a photon and a phonon. The probability of the inverse process depends on the number of phonons in the lattice. Therefore, by cooling the crystal, i.e., by reducing the number of phonons, we can make this probability as low as we please. Thus also the absorption of radiation at sufficiently low temperatures becomes weak. At the same time, radiative transitions representing pair recombination have finite probability even in the complete absence of phonons, i.e., at absolute zero. One would expect that at sufficiently low temperatures even a slight increase of the carrier density compared with the equilibrium density could make the probability of stimulated emission of photons greater than the probability of their absorption in the inverse transition.

The system of transitions can be represented by the three-level system shown in Fig. 64. Level 1 represents the ground state of the crystal, i.e., the absence of phonons and electron-

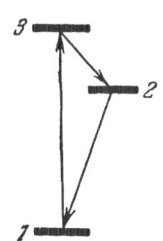

Fig. 64. System of indirect transitions in a semiconductor considered as a three-level system.

hole pairs.* Level 2 represents the presence of a phonon of energy $\hbar\Omega$; and level 3 represents the presence of a nonequilibrium carrier pair. The emission of a photon of energy $\hbar\omega$ is represented by a transition from level 3 to level 2, the act of emission leaving the lattice excited (a phonon $\hbar\Omega$ is generated). The probability of an inverse transition from level 2 to level 3 is proportional to the probability of the presence of a phonon in the lattice, which is equal to $\exp(-\hbar\Omega/kT)$ and becomes small as $T \rightarrow 0$. If exciton states exist in a semiconductor, the radiative recombination of carriers in excitons is more likely than the radiative recombination of free carriers [39]. The data of Haynes et al., who investigated the recombination radiation of silicon at low temperatures, and the results of Davies [40] indicate that recombination preceded by the formation of excitons is very likely.

The considerations applicable to population inversion in the case of indirect electron transitions may be extended to indirect recombination radiation of an electron and a hole bound in the form of an exciton.

We have pointed out that population inversion is the necessary but not sufficient condition for the amplification or generation of coherent electromagnetic radiation. To achieve population inversion (negative absorption), the probability of stimulated emission must be greater than that of absorption by free carriers.

The probability of stimulated emission by indirect transitions increases as the square of the carrier density but the probability of absorption by carriers is a linear function of the carrier density. Therefore, at some excitation levels the stimulated emission process should predominate.

Practical experience with the first masers based on the use of luminescent crystals (ruby and calcium fluoride containing uranium atoms) has shown that considerable difficulties arise, some of which have just been referred to. One of the problems is the preparation of materials in which carriers have sufficiently

*It is assumed that there are no impurity centers or structural defects.

long lifetimes at low temperatures and at carrier densities which depart strongly from the equilibrium density. At present, even in the purest semiconductor crystals, the lifetimes are governed by the recombination at impurity centers and defects (Chap. II). Moreover, at high nonequilibrium carrier densities, impact (triple) recombination may be active. The special characteristics of the recombination at low temperatures have not yet been extensively studied.

Further investigation of the band structure and impurity level spectra of semiconductors will undoubtedly lead to new ways of achieving stimulated emission. Intensive work on these subjects and important breakthroughs in the analysis of the band structure and in the control of the impurity content leads us to expect, in the immediate future, semiconductor masers working in the infrared region which will find important applications in science and technology.

Chapter V

CHANGES IN THE PROPERTIES OF SEMICONDUCTORS DUE TO BOMBARDMENT WITH FAST ELECTRONS, GAMMA RAYS, NEUTRONS, AND HEAVY CHARGED PARTICLES

The problem of the nature of changes in the structure and properties of solids subjected to penetrating radiation has become particularly important since the construction of the first nuclear reactors. Somewhat later, beginning around 1950, studies of the processes of radiation-defect formation and of the influence of these defects on the properties of semiconductors have attracted the attention of many investigators in connection with the further development of semiconductor physics and availability of perfect single crystals. At present, many important projects in this field are being undertaken in various laboratories in the USSR, USA, France, and elsewhere, and the number of papers published on this subject runs into hundreds. The theory of the formation of radiation defects and the experimental methods of investigating crystals containing such defects are being constantly improved.

The first part of the present chapter (Part A) presents briefly the fundamentals of the theory of radiation-defect formation by fast electrons, gamma rays, heavy charged particles and neutrons. The second part (B) reviews the currently available experimental data on the influence of radiation defects on the electrical conductivity, on the recombination of nonequilibrium carriers, and on the optical properties of semiconductors, particularly silicon and germanium.

PART A. THEORETICAL INTERPRETATIONS OF THE PROCESS OF RADIATION-DEFECT FORMATION

§23. Effects of Fast Electrons and Gamma Rays

A. Initial Assumptions

In the theory of radiation defects in solids, it is usually assumed that the simplest type of defect is an empty lattice site (a vacancy) and an atom occupying more or less stable interstitial position. Such defects are known as Frenkel defects. They belong to the class of point defects, which are simpler than dislocations and other more complicated and larger structural defects.

Another assumption is the existence of a "threshold energy" E_d which must be given to an atom in order to displace it to an interstitial position and thus produce a defect. Assuming the "impact" formation of radiation defects, Seitz [1] concluded that the value of E_d was several times greater than the energy necessary for the adiabatic displacement of an atom from its normal site to an interstice; he showed that the probable value of E_d was 25 eV for crystals with an atomic binding energy close to 10 eV.

Following this assumption, one would expect electrons of relatively low energy (hundreds of kiloelectron-volts) to be scattered on the atoms of a crystal and to form Frenkel defects. The high sensitivity to structural defects of such semiconductors as germanium and silicon has caused experimenters to study threshold energies and the influence of relatively simple radiation defects on semiconductors.

We shall show that by using the theory of electron scattering on nuclei of crystal atoms, and taking into account the symmetry and binding energy of the atoms, one can give a better description of the process of Frenkel defect formation by fast electrons.

Let us assume that a fast electron moves along the z axis (Fig. 65) and that its energy and momentum are E and P_e, respectively. We shall consider the collision of this electron with an atom at rest, assuming that after the collision the electron moves near the point O in the plane xOz at angle θ_e to the z axis, and that its new momentum is P_e'. After the collision, the atom also moves in the plane xOz at an angle θ_A to the z axis and its momentum is P_A. From the law of conservation of momentum, it follows that

$$\left.\begin{aligned}
P_e &= P_e' \cos\theta_e + P_A \cos\theta_A, \\
0 &= P_e' \sin\theta_e - P_A \sin\theta_A, \\
P_A^2 &= (P_e - P_e' \cos\theta_e)^2 + P_e'^2 \sin^2\theta_e, \\
\tan\theta_A &= \frac{P_e' \sin\theta_e}{P_e - P_e' \cos\theta_e}.
\end{aligned}\right\} \qquad (5.1)$$

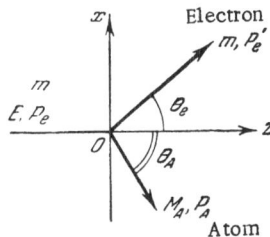

Fig. 65. Scattering of a fast
electron.

The atomic masses M_A are much greater than the electron mass; for example,

$$M_{Si} \approx 5 \cdot 10^4 m;$$
$$M_{Ge} \approx 1.3 \cdot 10^5 m.$$

Thus, $P_e' \approx P_e$, and, consequently,

$$P_A^2 = P_e^2 \sin^2\left(\frac{\theta_e}{2}\right);$$
$$\theta_A = \frac{\pi}{2} - \frac{\theta_e}{2}. \qquad (5.2)$$

We shall denote the kinetic energy of the atom after the collision by $E_A = P_A^2/2M_A$. Using $mc^2 = 0.511$ MeV $= E_0$ (the rest energy of the electron), we can write:

$$E_A = E_0 \frac{2m}{M_A}\left[\left(\frac{E}{E_0}\right)^2 + 2\frac{E}{E_0}\right]\cos^2\theta_A = E_{A\,max}\cos^2\theta_A. \qquad (5.3)$$

The maximum energy transferred to Ge and Si atoms is given in Table 9 for several values of the incident electron energy ($\theta_A = 0$, $\theta_e = \pi$).

B. Angular and Energy Distribution of Atoms after Electron Scattering

We shall denote by $\sigma(\theta_A)$ the cross section representing the probability that the momentum transferred to an atom is directed at an angle θ_A to the z axis.

Let the differential cross section for electron scattering at an angle θ_e be $\sigma_e(\theta_e)$. Then

$$\sigma_e(\theta_e)\sin\theta_e\,d\theta_e = \sigma(\theta_A)\sin\theta_A\,d\theta_A, \qquad (5.4)$$

where θ_e and θ_A are related by $\theta_A = (\pi/2) - (\theta_e/2)$. It follows from Eqs. (5.2) and (5.4) that
when $\theta_A < \pi/2$

$$\sigma(\theta_A) = 4\sigma_e(\pi - 2\theta_A)\cos\theta_A, \qquad (5.5)$$

TABLE 9

Electron energy E, MeV	$E_{Ge\ max}$, eV	$E_{Si\ max}$, eV	Electron energy E, MeV	$E_{Ge\ max}$, eV	$E_{Si\ max}$, eV
0.01	0.3	0.8	1.0	59	152
0.1	3.2	3.2	1.5	111	285
0.2	7.4	19	2.0	177	455
0.3	12	30.7	2.5	258	660
0.5	23	59	3.0	354	920

when $\theta_A > \pi/2$

$$\sigma(\theta_A) = 0. \tag{5.6}$$

Seitz and Koehler [2] have shown that the screening of the atomic nucleus by the electron cloud is unimportant in the case of collisions of fast electrons and nuclei, which are accompanied by the knocking out of atoms from crystal lattice sites. Such collisions can be regarded as the scattering of a relativistic electron in a Coulomb field. In the first approximation, which is satisfactory in the case of low values of θ_e and low energies transferred to atoms, we can use for σ_e the relativistic Rutherford cross section

$$\sigma_R(\theta_e) = \sigma_0 \frac{1}{\sin^4 \frac{\theta_e}{2}}, \tag{5.7}$$

where

$$\sigma_0 = \left(\frac{Zq^2}{2mc^2}\right)^2 \frac{1-\beta^2}{\beta^4},$$

Z is the nuclear charge, and $\beta = v/c$.

For the Rutherford scattering cross section

$$\left.\begin{array}{l} \sigma(\theta_A) = 4\sigma_0 \dfrac{1}{\cos^3 \theta_A} \quad \text{when } \theta_A < \dfrac{\pi}{2}, \\[2mm] \sigma(\theta_A) = 0 \qquad\qquad \text{when } \theta_A > \dfrac{\pi}{2}. \end{array}\right\} \tag{5.8}$$

Thus, the cross section representing the probability of the atomic momentum being directed at the angle θ_A has a sharp maximum at $\theta_A \leq \pi/2$ close to $\theta_A = \pi/2$, when the "impact parameter" is large and the electron is scattered at a small angle. It follows from Eq. (5.8) that the energy transferred is small. In fact, the cross section for electron scattering $\sigma_e(\theta_e)$ deviates considerably from the Rutherford value. According to McKinley [3], the cross section $\sigma_e(\theta_e)$ for light elements can be written as follows

$$\sigma_e(\theta_e) = \sigma_R(\theta_e) \, B\left(\sin \frac{\theta_e}{2}\right), \tag{5.9}$$

where the function B has the following form:

$$B(x) = 1 - \beta^2 x^2 + \pi \alpha \beta x (1 - x), \tag{5.10}$$

$x = \sin(\theta_e/2)$, and $\alpha = Zq^2/hc$. For the cross section σ, we find the expression

$$\sigma(\theta_A) = 4\sigma_0 \frac{B(\cos \theta_A)}{\cos^3 \theta_A}. \tag{5.11}$$

An important point, which follows both from the approximate and rigorous treatments, is the tendency for the momentum to be transferred through a large angle to the direction of the incident electron. The higher the electron energy, the closer to $\theta_A = \pi/2$ is the grouping of the collisions accompanied by energy transfer above a certain minimum (threshold).

The energy spectrum of atoms * excited by fast-electron scattering can be calculated as follows. Let $n(E_A)dE_A$ be the number of atoms which have received energy between E_A and $E_A + dE_A$ from a unit electron beam. Then

$$n(E_A)\,dE_A = 2\pi\sigma_e(\theta_e)\sin\theta_e\,d\theta_e. \tag{5.12}$$

From the relationship between the angles θ_e and θ_A, it follows that

$$\left.\begin{aligned}
E_A &= E_{A\max}\sin^2\frac{\theta_e}{2} = E_{A\max}\cos^2\theta_A, \\
dE_A &= \frac{1}{2}E_{A\max}\sin\theta_A\,d\theta_A.
\end{aligned}\right\} \tag{5.13}$$

* We mean here the atoms which have acquired excess kinetic energy.

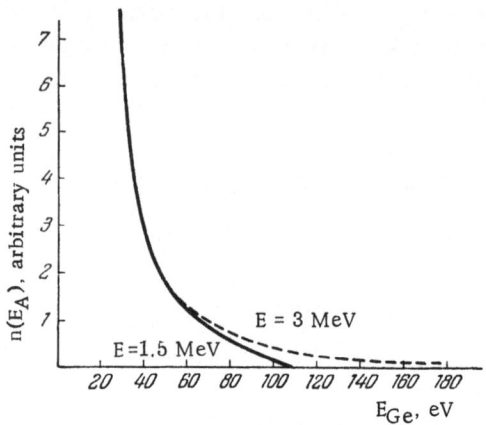

Fig. 66. Distribution of energies transferred to germanium atoms by electrons of incident energies
E = 1.5 MeV and 3 MeV.

Hence, we can find the relationship between the energy spectrum
and the differential cross section for electron scattering:

$$n\left(E_A\right) = \frac{4\pi}{E_{A\,\text{max}}}\, \sigma_e\, 2\sin^{-1}\left[\left(\frac{E_A}{E_{A\,\text{max}}}\right)^{1/2}\right]. \qquad (5.14)$$

We shall use the Rutherford formula and find that

$$n\left(E_A\right) = \frac{4\pi\sigma_0}{E_{A\,\text{max}}}\left(\frac{E_{A\,\text{max}}}{E_A}\right)^2. \qquad (5.15)$$

It follows from the above expression that most of the atoms have
low energies. The nature of this dependence is retained when the
exact scattering cross section formula is used. Figure 66 shows
the distribution of energies transferred to germanium atoms by
electron bombardment. Except for the region corresponding to
near head-on collisions, the curves are identical for two different
energies. It follows that the majority of defects generated by the
bombardment of crystals with 1.5 or 3 MeV electrons are due to
collisions in which the atoms receive not more than 50-70 eV.

C. Probability of the Formation of Frenkel Defects

Making the simplest assumption that the threshold energy
E_d needed for the displacement of an atom into an interstitial

position, is independent of the direction of the atomic momentum, and that each scattering act, accompanied by the transfer of energy $E_A > E_d$, produces one Frenkel defect,* it is found that the above formulas allow the calculation of the cross section Σ_d, which represents the probability of defect formation. Since the probability of defect formation depends on the electron energy E, the expression which will be given below is valid only for low-energy losses ΔE by electrons ($\Delta E \ll E$), i.e., for sufficiently thin samples. In order to calculate N_d, the concentration of defects produced by an integral electron flux Φ per 1 cm^2, the value of $\Sigma_d(E)$ must be multiplied by Φ and by the number of atoms in 1 cm^3 (N):

$$N_d = \Sigma_d(E)\,\Phi N. \tag{5.16}$$

Obviously, defects would be formed in all collisions for which the angle θ_A lies between $\theta_A = 0$ (head-on collision) and $\theta_{A\,max}$, which is found from the condition $E_{A\,max} \cos^2\theta_{A\,max} = E_d$, i.e.,

$$\cos\theta_{A\,max} = \sqrt{\frac{E_d}{E_{A\,max}}}. \tag{5.17}$$

The final expression for $\Sigma_d(E)$ has the form:†

$$\Sigma_d(E) = 8\pi\sigma_0 \left[\frac{1}{2}\left(\frac{1}{x_0^2} - 1 \right) + \pi\alpha\beta\left(\frac{1}{x_0} - 1 \right) + (\beta^2 + \pi\alpha\beta)\ln x_0 \right], \tag{5.18}$$

where

$$\sigma_0 = \left(\frac{Zq^2}{2mc^2} \right)^2 \frac{1 - \beta^2}{\beta^4},$$

$$\alpha = \frac{Zq^2}{hc} = 0.23 \ \text{for} \ \ \text{Ge and } 0.1 \ \text{for} \ \ \text{Si},$$

$$x_0 = \cos\theta_{A\,max} = \sqrt{\frac{E_d}{E_{A\,max}}},$$

$$\beta = \frac{\left[\dfrac{2E}{mc^2} + \left(\dfrac{E}{mc^2} \right)^2 \right]^{1/2}}{1 + \dfrac{2E}{mc^2} + \left(\dfrac{E}{mc^2} \right)^2}.$$

* This assumption is valid for "threshold" experiments when the value of E is only slightly greater than the threshold energy (for example, when E < 1 MeV in experiments on germanium).

† The expression given here is identical with the formula of Lark-Horovitz [4], who used a function of the angle θ_e.

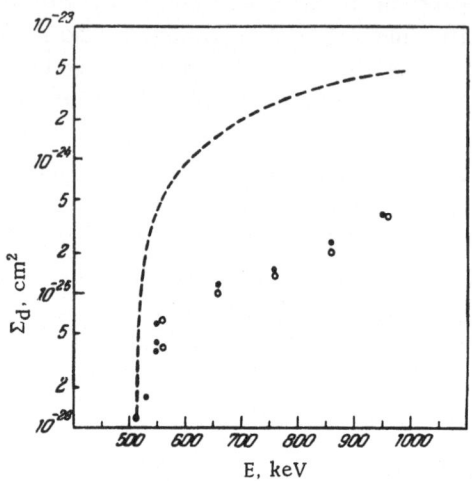

Fig. 67. Energy dependence of the cross section
for Frenkel defect formation in germanium, calculated on the assumption that E_{min} = 0.51 MeV
(dashed curve).

Thus, the function $\Sigma_d(E)$ for a crystal consisting of identical atoms
is governed by the parameter E_d. The form of the function $\Sigma_d(E)$
for germanium is given in Fig. 67.

In some cases, it is important to know the total number of
atoms knocked out by fast-electron bombardment. Using the foregoing assumptions on the existence of the threshold energy E_d,
which is independent of the direction of electron incidence, Kahn
[5] calculated the number of atoms $N^+(E)$ knocked out by a single
incident electron in germanium and silicon. Kahn's dependences
of the value of N^+ on the electron energy E are given in Fig. 68.
Figure 69 presents the range-energy relationship for electrons
[6]. It should be remembered that the ordinate axis gives the
range along the trajectory and does not take into account the scatter of the electron directions. In calculating the number of knocked-
out atoms, Kahn assumed that if the energy of the primary knocked-
out atom E_A is greater than $2E_d$, secondary and other defects are
formed and the total number of defects ν for such collisions is
$\nu = E_A/2E_d$, in accordance with the cascade theory of Kinchin
and Pease [7] (cf. Sec. 27).

The theory of the scattering of relativistic electrons by nuclei, developed by Mott, McKinley, and Feshbach was checked

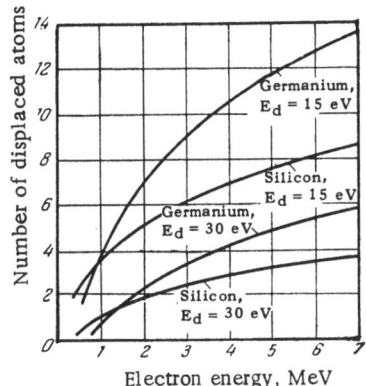

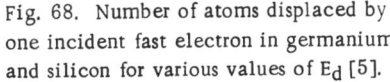

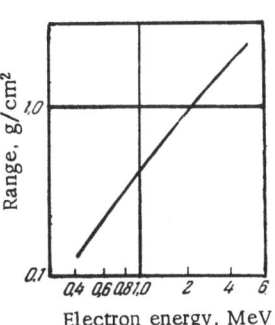

Fig. 68. Number of atoms displaced by
one incident fast electron in germanium
and silicon for various values of E_d [5].

Fig. 69. Range-energy relationship for electrons.

by investigating scattering in thin metal foils; the agreement between the theory and experiment was better than 1% [8]. Therefore, the nature of the primary act of energy transfer from fast
electrons to atoms in crystals can be regarded as reliably established, in contrast to the later stages of the formation and
stabilization of defects.

The forces of interaction between an atom which has received momentum and neighboring atoms in a lattice have practically no influence on the initial motion of this atom (in the later
stages, these forces govern the dissipation of the energy and the
formation of defects). The "collision time" τ_i for electrons of
kinetic energies of several MeV is

$$\tau_i \approx \frac{\lambda}{c} \approx 10^{-20} \text{ sec.}$$

(Here $\lambda = h/mc$ is the Compton wavelength.) This time is very
short compared with that required for the displacement of an atom
to a typical interstitial position: the initial velocity of the displaced atom is of the order of 10^6 cm/sec, assuming that all its
energy is kinetic. Thus, the order of magnitude of the time for
a displacement of an atom by 10^{-8} cm is

$$\tau_{\text{displ}} \approx \frac{10^{-8}}{10^6} = 10^{-14} \text{ sec.}$$

§24. Improved Theory of the Displacement
of an Atom from Its Site to an Interstice
in a Diamond-Type Crystal Lattice
(Germanium)

After the collision of an electron with an atom, the forces acting on the displaced atom tend to make it return to its initial position. For small displacements, the forces may be estimated from the well-known values of the elastic constants. In the present chapter, we shall consider only germanium. An atom of germanium with a kinetic energy of 25 eV has a velocity of the order of 10^6 cm/sec, which is very small compared with the "velocities" of electrons – as represented by their binding energy in a crystal – which reach values of about 10^8 cm/sec. At any given moment, electrons are in states corresponding to the lowest energy of a given configuration of nuclei, as in the case of the elastic deformation of a crystal [9]. We shall use the results of H. Smith's theory of the elastic constants of diamond-type crystals [10], which is based on some simplifying assumptions about the forces of interaction between the nearest atoms. According to this theory, the energy ΔE_A which is needed to displace an atom in any direction by a small distance x, may be expressed as follows

$$\Delta E_A = c \left(\frac{x}{b} \right)^2, \qquad (5.19)$$

where the value of c for germanium is 18.3 eV, and b = 1.41 × 10^{-8} cm is the distance between two nearest-neighbor atoms. Using the expression (5.19), the data on the presence of "natural" interstitial positions in diamond-type lattices, and the well-known values of the atomic binding energies in crystals, we can estimate the energy needed to displace an atom from its regular site to the nearest interstitial position. It is understood that this energy estimate is not accurate since the atomic displacements cannot be regarded as small.

Let us assume that an atom O, which has received kinetic energy as a result of the scattering of a fast electron lies at the origin of coordinates, and that the four nearest atoms A, B, C, D are at the points

$$A - b(1, 1, 1), \qquad B - b(1, -1, 1),$$
$$C - b(1, 1, -1), \qquad D - b(-1, -1, 1)$$

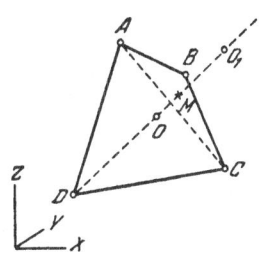

Fig. 70. Positions of nearest neighbors and natural interstitial positions in a diamond-type lattice.

(cf. Fig. 70). One of the nearest interstitial positions O_1 is the point $b(1,1,1)$ at a distance OO_1, equal to $b\sqrt{3} \approx 1.73b$.

The distance d from O to M at $b(1/3, 1/3, 1/3)$, which is at the intersection of the line OO_1 and the plane in which lie the three nearest atoms, is $d \approx 0.6b$. The potential energy $\Delta E_A(OM)$, corresponding to the displacement d_{OM}, amounts to about 6 eV. Kohn's detailed analysis of the interaction of the atom at O with its neighbors, when the former is displaced in the direction OO_1 or in the directions of other nearest interstitial positions [(0, 0, 2), (0, 2, 0), or (2, 0, 0)], shows that the energy representing displacements of the nearest atoms is small compared with $\Delta E_A(OM)$.

A free germanium (or silicon) atom has the external electron configuration $(4S)^2(4P)^2$. In a crystal lattice, the probability of a configuration of outer electrons is given by the combination of the wave functions of the S- and P states. The maximum probability of finding an electron is along the lines joining neighboring atoms. We may assume approximately that the "covalent binding energy" can be calculated by dividing the binding energy of the crystal by four times the number of atoms. The experimentally determined crystal binding energy of germanium is $W_{Ge} = 89$ cal/mole. It follows that the "covalent binding energy" amounts to $W_{Ge}/4N_{Ge} \approx 1.92$ eV.

We have discussed the process of primary energy transfer to one of the atoms in a crystal, and we have estimated the change in the potential energy of the atoms caused by the displacement. Using these data we can obtain quantitative estimates of the threshold energy E_d which is needed to transfer an atom to one of the nearest interstitial positions. *

In a diamond-type crystal, an atom at a site O has the following nearest "natural" interstitial positions:

$$M_1 - b(1, 1, 1), \qquad M_2 - b(2, 0, 0),$$
$$M_3 - b(3, 1, -1), \qquad M_4 - b(2, 2, 2).$$

* The experimental data for germanium and silicon show that some of these theoretical conclusions are justified. However, the idea that the probability of the formation of a stable defect is unity if the energy is higher than the threshold value ($E_A > E_d$) is not justified for germanium or silicon, at least when the crystal temperature is not too low.

The minimum energy for the displacement of this atom, and its stability in the displaced position, need not be the same for each of these four positions [11].

Displacement $O \rightarrow M_1$. The displacement from O to M_1 increases the potential energy by 6 eV, but the energies of the remaining atoms remain practically constant. Estimating the upper limit of the value of E_d for the displacement OM_1, i.e., along the [111] axis, we add the potential energy $\Delta E_A (OM_1)$ and the energy necessary to break four covalent bonds:

$$E_d (OM_1) = (6 + 4 \cdot 1.92) \text{ ev} = 13.7 \text{ ev}.$$

This value is very close to the experimental values of E_d obtained in the recent work of Brown and Augustyniak [12], and Smirnov [13]. Brown and Augustyniak found that E_d amounted to 15.3 eV; Smirnov reported a value close to 15.7 eV. However, the theory shows that the potential barrier is not of sufficient height to prevent the atom from returning to O, and therefore the defect is not very stable.

Displacement $O \rightarrow M_2$, i.e., to the points $b(2, 0, 0)$, $b(0,2,0)$, or $b(0, 0, 2)$. In this case, in contrast to the displacement OM, the atom O should pass through a point lying between the atoms B and C (Fig. 70). Using the expressions for the potential energy of the displacement, and allowing for the interactions with B and C and the need to break covalent bonds, it is found that the threshold energy is higher than in the preceeding case:

$$E_d (OM_2) = [33.5 \pm 4] \text{ ev}.$$

An energy of the order of 3-4 eV is needed to displace the atoms B and C. When B and C return to their initial positions, they stabilize the defect by forming a potential barrier.

Displacements $O \rightarrow M_3$ and $O \rightarrow M_4$ are, according to the model used, accompanied by the formation of relatively unstable defects, since potential barriers are not formed after these displacements.

The remaining natural interstitial positions surrounding the atom O may be occupied by this atom only after collisions with neighboring atoms, which are accompanied by considerable energy transfer; consequently, the values of E_d are, in these cases, considerably greater than that for the displacements $O \rightarrow M_1$ or $O \rightarrow M_2$.

The example considered shows that the threshold energies should be close to the values suggested by F. Seitz. Moreover, one would expect that in diamond-type lattices near the defect formation threshold the number of defects generated would depend strongly on the mutual orientations of the crystal axes and the direction of the incident electron beam.

§ 25. Formation of Radiation Defects
by Gamma Rays

Gamma rays, like fast electrons, produce structural defects of the "point" type. Galavanov [14] and Kahn [5] showed that the probability of displacing atoms by the direct interaction of gamma quanta with nuclei is very small. The main effect is due to fast electrons, produced by the photoeffect and the Compton effect, and of electron-positron pairs produced by high-energy gamma rays. The total cross section for the absorption of gamma rays μ is determined by the three processes referred to above. Figures 71 and 72 show the total cross section for the absorption of gamma rays in silicon and germanium as a function of the photon energy.

To calculate the concentration of the knocked-out atoms, it is necessary to invoke the theory of the formation of Frenkel defects by fast electrons and positrons * (discussed earlier). Calculations of the number of atoms displaced as a indirect result gamma-rays bombardment of silicon and germanium have been carried out by Kahn [5]. In these calculations he made the usual assumption of the existence of a definite threshold energy for the appearance of defects, E_d, and of a unique value of the probability of defect formation when $E_A > E_d$. The results of Kahn's calculations are shown in Figs. 73 and 74.

Similar data for other elements are given in [15], and in the book by Dienes and Vineyard [16].

Thus, the problem of the effect of gamma rays on semiconductors, which is of great practical importance, reduces to a calculation of the probability of the generation of fast electrons or positrons in a crystal, and to a study of the influence of the re-

---

*In the case of positrons, we must allow for the probability of annihilation, which reduces somewhat the number of defects produced by positrons.

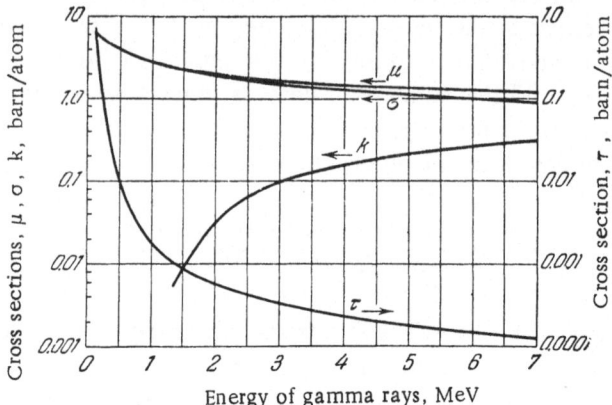

Fig. 71. Cross section for the absorption of gamma rays in silicon: τ represents the photoeffect; σ, the Compton effect; k, the formation of electron-positron pairs; μ, the total cross section.

sultant defects on the properties of a semiconductor. Judging by the available data, the properties (for example, the energy level spectrum) of defects produced by fast electrons and gamma rays are identical, in agreement with the theory.

§ 26. Effects of Fast Neutrons and Heavy Charged Particles

A. Energy of Recoil Nuclei

Fast electrons give rise to structural defects by transferring some of their kinetic energy to atomic nuclei. Usually, the recoil nucleus drags behind it its electron shell. F. Seitz and J. Dienes have shown that (see foregoing section) the outer-shell electrons, whose binding to the atoms is weak, may be relatively easily detached from the atom whose nucleus has been hit by a neutron. The most probable process is the elastic scattering of the fast neutrons. The energy transferred to the nucleus varies from zero to

$$E_{A\max} = \frac{4M_n M_A}{(M_n + M_A)^2} E_n, \qquad (5.20)$$

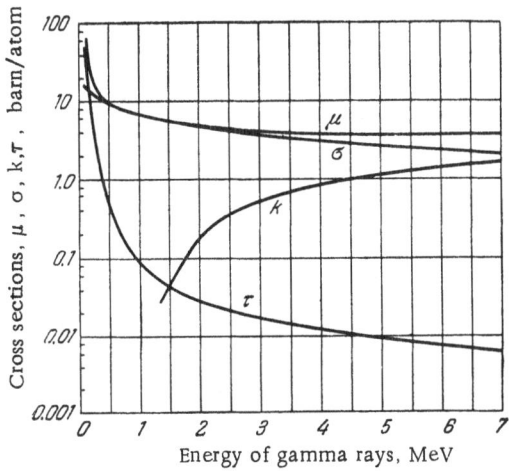

Fig. 72. Cross section for the absorption of gamma rays
in germanium: τ represents the photoeffect; σ, the
Compton effect; k, the formation of electron-positron
pairs; μ, the total cross section.

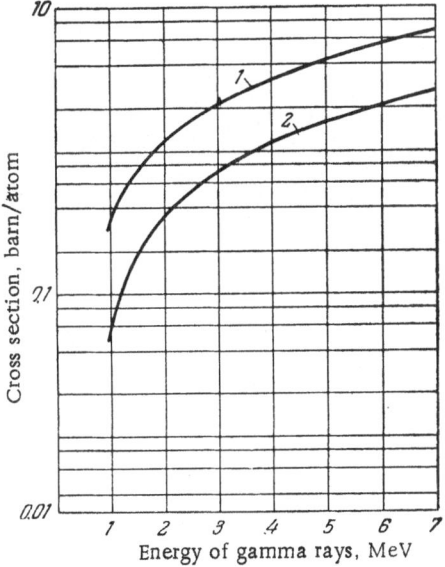

Fig. 73. Cross section S_d representing the probability of knocking out silicon atoms as a result of
irradiation with gamma rays, assuming that E_d = 15 eV
(1) and 30 eV (2).

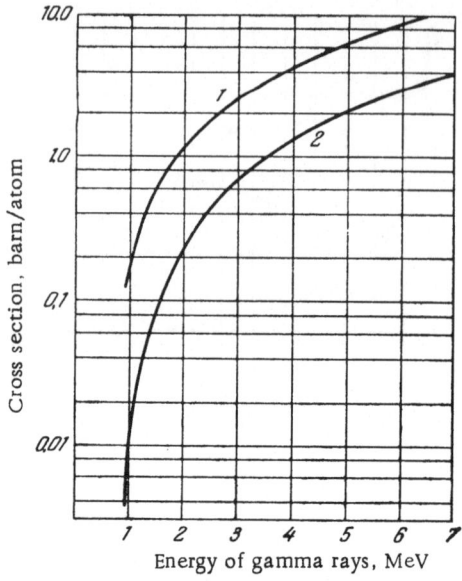

Fig. 74. Cross section S_d representing the probability of knocking out germanium atoms as a result of irradiation with gamma rays, assuming that E_d = 15 eV (1) and 30 eV (2).

where M_n is the mass and E_n is the kinetic energy of a neutron. The distribution of the recoil-atom energies is related to the angular distribution of the scattered neutrons. The simplest assumption, which is frequently used, is that the neutron scattering is isotropic. All the values of the recoil energy from 0 to $E_{A \max}$ are then equally probable. The differential cross section $d\sigma$ for the energy transfer in the region E_A, $E_A + dE_A$ is

$$d\sigma = \frac{\sigma_t}{E_{A \max}} dE_{A}, \qquad (5.21)$$

where σ_t is the total neutron cross section, which is assumed to be wholly due to the elastic scattering. The average energy transferred on scattering is

$$\overline{E}_A = \frac{1}{2} E_{A \max}. \qquad (5.22)$$

The value of σ_t lies usually within the range 1-10 barns for fast neutrons (produced by fission). These neutrons have energies up

to 15 MeV, the average energy being of the order of 2 MeV. If
we assume that the average neutron energy is indeed 2 MeV, * it
follows from Eqs. (5.20) and (5.22) that the average energy of recoil nuclei (atoms) is

$$\overline{E}_A = \frac{4}{A}\left(1 + \frac{1}{A}\right)^{-2} \approx \frac{4}{A} \text{ MeV,} \qquad (5.23)$$

where A is the atomic weight.

The average energy of recoil atoms resulting from bombardment with fast neutrons is many times greater than the average
energy of recoil atoms resulting from bombardment with heavy
charged particles (α particles, protons) of the same energy as the
neutrons; this is a consequence of the difference in the scattering
processes.

It has been shown recently, both experimentally and theoretically, that the assumption of isotropic scattering of fast neutrons
is not a good approximation. In fact, neutrons of energies of the
order of 1 MeV are scattered primarily in the forward direction.
Consequently, the average transferred energy is less than that
calculated using Eq. (5.22). Inelastic scattering is also possible,
and this should reduce the average kinetic energy of recoil nuclei.
Correction coefficients f, which should be used to multiply the
"isotropic" value of $\overline{E}_A$ in order to obtain the correct value, have
been given in [16], and are reproduced below:

Element	Be	C	Al	Cr	Fe	Ni	Cu
Value of f	0.56	0.84	0.58	0.57	0.57	0.64	0.60

The anisotropy of the scattering lowers the value of $\overline{E}_A$ in
most elements by 30-50% for neutrons in the range 1-2 MeV. At
higher neutron energies, this correction is greater.

B.   Formation of Secondary Structural Defects
as a Result of Elastic Collision Cascades

We have shown above that the kinetic energy transferred by
fast neutrons to nuclei in a crystal considerably exceeds the "thresh-

* When the irradiation is carried out in a reactor, the spectrum of neutrons incident on
a sample may be essentially different from the spectrum of "primary" neutrons because of the experimental conditions, especially the geometry and the material used
as a moderator.

hold" for the formation of defects, E_d. * The primary knocked-out atoms are, in turn, capable of knocking out other atoms, so that the final total number of structural defects is always con - siderably greater than the number of collisions between the in- cident fast neutrons and atoms in the crystal. The number of atoms knocked out in the primary collision, taken per unit vol- ume, is

$$N_p = \Phi t N_0 \sigma_d, \qquad (5.24)$$

where Φ is the number of bombarding particles crossing an area 1 cm^2 in 1 sec (the flux density), t is the duration of bombard- ment, N_0 is the number of atoms in 1 cm^3 of the substance, and σ_d is the cross section for collisions which give rise to the pri- mary knocked-out atoms.† In the case of neutron bombardment, it is usually assumed that $\sigma_d = \sigma_t$. We shall use ν to denote the number of atoms displaced by the primary atom, including the primary atom itself. The value of ν depends on the energy of the primary atom; averaging over energies gives $\bar{\nu}$. The total num- ber of displaced atoms in 1 cm^3, N_d, is given by

$$N_d = \bar{\nu} N_p. \qquad (5.25)$$

We shall assume that all collisions are of the pair type. This assumption is justified by the fact that the radius of the sphere of interaction of the forces is considerably smaller than the inter- atomic spacings in the crystal. We shall also assume that the atoms in the crystal lattice are initially at rest; we are not taking into account the ordering of atoms. These assumptions are made in all calculations of the cascade multiplication of defects in crys- tals. Moreover, some of the workers [17, 2, 18, 7] made addi- tional assumptions and, depending on the simplifications intro- duced, obtained slightly different quantitative results. Kinchin

---

* This statement is also valid for bombardment with heavy charged particles of energies in the region of several MeV. The transfer of energy from heavy charged particles to atoms in a crystal has been discussed by several workers [2, 4, 16]; since the secondary processes of defect "multiplication" depend only on the energy transferred to a sub- stance, the theory considered above can also be used to interpret the results of bom- bardment with α particles, deuterons, protons, etc.

† The term "primary atom" will be used to denote an atom knocked out of its site and having considerable kinetic energy.

and Pease [7], whose results are frequently used to interpret the experimental data, assumed that:

a. the primary atom loses energy only by ionization until its kinetic energy drops to the threshold value E_I which is given, according to Seitz, by

$$E_I = \frac{1}{8} \frac{M_1}{m} E_g, \tag{5.26}$$

where M_1 is the mass of the moving atom, m is the electron mass, E_g is the lowest electron-excitation energy, which is identical with the "optical" forbidden bandwidth;

b. all the moving atoms having energies less than E_I lose energy solely by elastic collisions with the atoms at rest;

c. an atom always leaves its site if collision with another atom gives it a kinetic energy E_A which is greater than the threshold energy E_d, but it remains at its site if $E_A < E_d$;

d. the incident atom remains at the site of an atom at rest if it transfers to the latter an energy greater than E_d so that the incident atom energy after the collision is $E_A < E_d$; the total number of defects increases only if, after the collision, both atoms have energies greater than E_d.

According to the model considered here, atoms do not overcome a potential barrier before colliding with other atoms and displacing them. Every atom leaves its site retaining the total kinetic energy received in a collision.

On the other hand, Seitz and Koehler [2], and Snyder and Neufeld [17] assumed that an atom lost a fraction of its kinetic energy, equal to E_d, before being able to knock-out other atoms. They assumed also that the incident atom could not remain at a vacated site. These differences in the assumptions almost compensate one another in quantitative calculations. In analyzing the cascade process, one normally uses the laws for the collision of hard spheres; Dienes and Vineyard [19] showed that the use of Rutherford's law, describing more accurately the high-energy collisions, did not greatly alter the calculated number of knocked-out atoms.

The dependence of the average number of defects ν on E_A * may be calculated as follows. Let us assume that E is less than

---

* We shall omit the subscript A from now on.

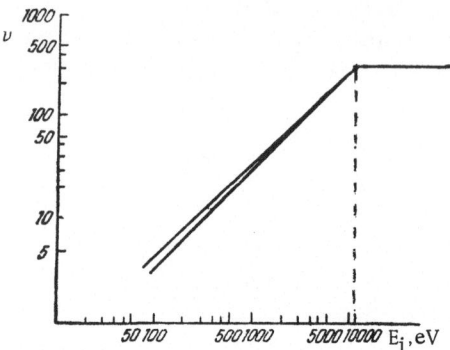

Fig. 75. Dependence of the number of knocked-out atoms ν on the primary atom energy (cascade theories) for germanium; E_d = 15 eV, E_i ≈ 12,000 eV.

E_i and greater than $2E_d$. On making an elastic collision, one atom transfers to the other an energy E_2' while retaining an energy E_1', where $E_1' + E_2' = E_1$. It follows from the law of hard-sphere collisions that all values of the transferred energy, from 0 to E, are equally likely, and the differential cross section $d\sigma$ for energy transfer in the energy range (E, E +dE) is

$$d\sigma = c' \, dE, \quad \text{where } c' = \frac{\pi a_1^2}{E_1}; \tag{5.27}$$

here, a_1 is the atomic "radius."

After the first collision, the number of further collisions which the first atom may undergo is $\nu(E_1')$ if $E'_1 \geq E_d$. The average number of displacements caused by the first atom can be represented in the form

$$\int_{E_d}^{E_1} \frac{1}{E_1} \nu(E_1') \, dE_1'. \tag{5.28}$$

The number of displacements caused by the second atom is $\nu(E_2')$ if $E_2' \geq E_d$, or zero if $E_2' < E_d$. Multiplying this number by the probability of a given energy distribution, and integrating over the energies E_2', we obtain a formula similar to Eq. (5.28). Adding the two integrals we find the fundamental equation for $\nu(E)$, which is valid in the range $2E_d \leq E \leq E_i$,

$$v(E) = \frac{2}{E} \int_{E_d}^{E} v(E') \, dE'. \tag{5.29}$$

Multiplying both sides of the above expression by E and differentiating with respect to E, we obtain the expression

$$E \frac{dv}{dE} = v \quad \text{(in the range } 2E_d < E < E_i), \tag{5.30}$$

the solution of which is $v(E) = CE$.

The constant C is found from the condition $v(2E_d) = 1$ as follows:

$$\left. \begin{array}{l} v(E) = 1 \text{ when } 0 < E < 2E_d, \\[2mm] v(E) = \dfrac{E}{2E_d} \text{ when } 2E_d < E < E_i, \\[2mm] v(E) = \dfrac{E_i}{2E_d} \text{ when } E > E_i. \end{array} \right\} \tag{5.31}$$

Thus, on the average, half the energy of the primary atom is used to form defects; the other half is lost in collisions which are not accompanied by atomic displacements.

Figure 75 shows the dependence $v(E)$, calculated in accordance with Kinchin's (lower curve) and Seitz's (upper curve) models. Near the threshold energy $(E \approx E_d)$ the difference between these two curves is considerable but at high energies the curves practically coincide. Seitz's results may be described quite well by the approximate formula

$$v(E) = 0.56 + 0.56 \frac{E}{E_d}. \tag{5.32}$$

To calculate the total number of defects in a crystal, the value of v must be averaged out over the spectrum of the primary atom energies. When substances of not too low atomic weight are bombarded with fast neutrons in a reactor, the majority of primary atoms have energies in the region where $v(E)$ may be regarded as a linear function of E. The averaging then leads to the replacement of E in the expression $v = E/2E_d$ with $\bar{E} = fE_{A\,max}/2$, i.e.,

$$\bar{v} = f \frac{2M_n M_A}{(M_n + M_A)^2} \frac{\bar{E}_n}{E_d}, \tag{5.33}$$

where $\bar{E}_n$ is the average kinetic energy of neutrons, and f is a factor which allows for the scattering anisotropy.

When light substances are irradiated or when high-energy neutrons are used, a considerable part of the energy of the primary atoms is lost by ionization, and the process of averaging becomes more complex.

When heavy charged particles are used in bombardment, the primary-atom energies are relatively low and ν (E) cannot be regarded as a linear function of the energy E. Using the equality

$$\bar{\nu} = \nu(\bar{E}), \qquad (5.34)$$

where the average energy transferred in the Rutherford scattering is

$$\bar{E} = \left(\frac{E_d E_{max}}{E_{max} - E_d} \right) \ln \frac{E_{max}}{E_d}, \qquad (5.35)$$

we can obtain a rough estimate of the value of $\bar{\nu}$. More exactly, the formula deduced by Dienes gives

$$\bar{\nu} = \frac{1}{2} \left(\frac{E_{max}}{E_{max} - E_d} \right) \left(1 + \ln \frac{E_{max}}{2E_d} \right). \qquad (5.36)$$

It is evident that bombardment with heavy particles of energies of the order of several MeV produces many fewer defects than bombardment with fast neutrons in a reactor. This is a direct consequence of the difference between the scattering in a Coulomb field and the scattering of neutrons by nuclei.

All the cascade models give only rough approximations and the results of calculations may differ by a factor of 2-3 from the true number of knocked-out atoms.

C. Collisions Involving the Displacement of Defects

Apart from the direct knocking out of atoms into interstitial positions, the irradiation of a crystal may be accompanied by the displacement of the generated defects by a chain-like transfer of momentum from moving atoms to atoms at rest.* Experimental proof that such processes occur is provided by the data on the disordering of alloys   which cannot be accounted for by the simple

---

* Sometimes processes of energy and momentum transfer along a chain of atoms are called "focused collisions."

cascade theory. A cascade-process theory modified to include collisions with substitution has been developed by Kinchin and Pease [7].

Kinchin and Pease assumed that the ionization and elastic collision regions are sharply divided and applied the laws of elastic collision between hard spheres. The following further assumptions were made:

a. an atom leaves its site when it receives a kinetic energy greater than E_d; moreover, if the energy transferred to this atom lies between E_r (the "substitution threshold energy") and E_d, where $E_r < E_d$, and the energy retained by the incident atom is less than E_d, the latter remains at the lattice site of the knocked-out atom, i.e., the defect moves to a new position;

b. the incident atom replaces the knocked-out atom if the latter receives an energy greater than E_d and the energy of the former is less than E_d after the collision.

As in the models discussed earlier, an increase in the number of defects is possible only if, after the collision, both atoms have energies greater than E_d. Therefore, the total number of defects remains constant. However, the calculated number of atom displacements in a lattice increases as a function of the selected value of E_r, which is the "substitution threshold energy." The method for calculating the number of such substitutions is similar to the method used earlier to calculate the number of defects $\nu (E)$. The average number of substitutions $\xi (E)$, caused by a primary atom with energy E, is given by

$$\xi(E) = \frac{E}{2E_d}\left(1.6 \ln \frac{E_d}{E_r} + 1\right) \text{ when } E \geqslant E_d. \tag{5.37}$$

Comparing the expressions for $\xi (E)$ and $\nu (E)$, we obtain the relationship

$$\frac{\xi (E)}{\nu (E)} = 1.6 \ln \frac{E_d}{E_r} + 1. \tag{5.38}$$

It is evident that if E_r is considerably smaller than E_d, the appearance of each new defect is accompanied by several substitutions. The value of E_r has not been estimated directly; Kinchin and Pease [7] assumed that $E_d \approx 10 E_r$ and therefore $\xi (E) \approx 5\nu (E)$. The allowance for substitution is important because the processes of defect healing by annealing depend strongly on the spatial separation of vacant sites and interstitial atoms.

D.   Thermal Theory of the Formation of Radiation Defect Aggregates

The considerable energy received by one atom in the primary scattering act (for example, in the scattering of neutrons) is distributed rapidly between a large number of atoms. The state of matter near the point of the initial collision can be represented approximately by the rapid heating of a small volume to a high temperature. It is very likely that, apart from the cascade multiplication of defects just considered, the final number of defects, which determines the change in the properties of the irradiated crystal, depends strongly on the rate of energy equalization in the strong excitation region near the site of the primary collision. Using the standard theory of thermal conduction, we can make some estimates about the heating process and develop a qualitative description of the processes which follow the energy transfer to the primary atom. We should remember, however, that the equilibrium distribution is reestablished rapidly in the excitation region and, therefore, strictly speaking, its state cannot be represented by some temperature. The times and distances, which are characteristic of the spread of the excitation region, are so small that the macroscopic laws of thermal conduction can be used only to provide a qualitative description. Seitz and Koehler [2] (cf. also [16]) are of the opinion that initially the excitation of matter may be so high that a large number of atoms in the excited region suffer a disordering, and a "displacement spike" is formed.

In terms of the thermal theory, the energy E_A transferred to an atom by an incident particle is assumed to be emitted suddenly in the form of thermal energy in a small volume of a continuous medium; this energy is assumed to be propagated in accordance with classical heat-conduction laws. The medium is represented by a thermal diffusivity D and a temperature $T(r, t)$ at every point r at a time t. The diffusivity D is related to the thermal conductivity C, the specific heat c, and the density d by

$$D = \frac{C}{cd} . \tag{5.39}$$

The thermal conductivity of semiconductors (and models of the same) is the sum of terms representing the thermal conductivity of the crystal lattice and the contribution of conduction-band electrons [20]. We may assume that the form of excitation considered here, related primarily to atom displacements, i.e., to the lattice, does not spread (at least in the initial stage) to charge carriers. Thus, the value of C should represent the thermal conductivity of the lattice. The value of D is usually close to 10^{-3} cm^2/sec [19]. The temperature in the excitation region obeys the heat-conduction law

$$\nabla^2 T = \frac{1}{D} \frac{\partial T}{\partial t}. \qquad (5.40)$$

The solution of the above equation for the case of the emission of an energy E_A at the origin of coordinates at a time $t = 0$, when the initial temperature of the substance is T_0 and r is the distance from the origin of coordinates, is given by

$$T(r, t) = T_0 + \frac{E_A}{(4\pi)^{3/2} cd} \frac{1}{(Dt)^{3/2}} e^{-\frac{r^2}{4Dt}}. \qquad (5.41)$$

At any given time the temperature is maximal near the origin of coordinates: in this region, the excess temperature is proportional to $t^{-3/2}$.

If the energy of the primary atom is high and the points at which it collides with other atoms are sufficiently far apart, the regions of thermal shock excitation may be regarded as spherical. As the primary atom slows down, these regions begin to overlap; we can consider the process as the uniform emission of thermal energy along the path of the primary atom. If the thermal energy per unit length of the track of a particle is Q' and ρ is the excitation region radius, the solution of the heat-conduction law is

$$T(\rho, t) = T_0 + \frac{Q'}{4\pi cd} \frac{1}{Dt} e^{-\frac{\rho^2}{4Dt}}. \qquad (5.42)$$

Numerical estimates of the rate of propagation of the shock-heated volume and of the probable temperatures are given for some metals in the work of Dienes and Vineyard [16].

The most radical approach to the process of defect formation in the case when primary-atom energies are high has been proposed by Brinkman [21]. According to his calculations, the average distance between collisions accompanied by the knocking out of secondary atoms becomes equal to the interatomic spacing at primary-atom energies as low as 2×10^4 eV. It follows that as soon as the primary atom energy decreases to this value, it is stopped very rapidly, producing a dense region of secondary collisions. It is not possible to consider separately each displacement (defect) within this region, i.e., a large number of atoms in an approximately cylindrical volume becomes completely disordered so that they resemble the molten state or the vapor of the substance. Brinkman assumed that near the track of the primary atom an "inversion" will occur, i.e., the atoms initially closest to the track will be knocked out furthest. The disordered region recrystallizes in 10^{-10}-10^{-12} sec, beginning from the outer boundary of the region. We may expect that the great majority of the atoms reestablish the initial lattice of the crystal. Unfortunately, Brinkman's discussion does not lead to any definite quantitative result. The initial assumptions of his theory (the frequency of collisions between atoms in the elastic interaction region is assumed to be higher than the values expected by other authors) were criticized by Seitz and Koehler [2]. Smirnov [78] has suggested that the "thermal" theory of the generation of radiation defects might be used also to deal with the bombardment of crystals by fast electrons, when the energy transferred to the primary atom E_A is only of the order of several tens of electron-volts.

Experimental checks of the concept of "thermal disordering" of relatively large regions in crystals bombarded with various particles can be obtained from the fact that the diffusion of impurities and self-diffusion are much easier in such regions.

PART B. RADIATION DEFECTS AND THE EFFECTS
OF NUCLEAR REACTIONS IN SEMICONDUCTORS
(EXPERIMENTAL DATA)

§ 27.   Purpose and Methods of Investigating
Radiation Defects in Semiconductors

The purpose of investigating structural defects in semiconductors is to obtain information on the general problems of solid-state physics, and to find new semiconducting materials with special properties governed primarily by the nature and energy level spectra of radiation defects and chemical impurities formed by nuclear reactions. *

* An interesting review of the theory aims, and methods of the study of radiation defects in solids and of the current state of this branch of physics was published by Brooks [22]. Brooks also reviewed topics outside the physics of semiconductors.

Semiconductors are especially convenient for investigating the threshold energies for defect formation, and for obtaining data on the defect generation process and the nature of simple and complex radiation defects.

Results important in the physics of real crystals are yielded by the study of the reestablishment of the equilibrium state in semiconductors as a result of the annealing of radiation defects. Furthermore, the study of defect stability is extremely important from the practical point of view. It is obvious that if defects give rise to new and desirable properties in a semiconductor (for example, the appearance of infrared photoconductivity), it is important to retain these properties throughout the service life of a device made from this semiconductor. On the other hand, the detrimental effects of radiation defects (for example, increase in the volume recombination velocity) can, in principle, be eliminated by selecting semiconducting materials in which defects are unstable or for which the threshold energies of defect formation are high.

Before considering the experimental data on radiation defects, let us deal briefly with some of the methods which have been found fruitful in the study of such defects.

To obtain information on the number of defects produced by hard radiation, and on the defect energy-level systems, the electrical conductivity and the Hall effect are usually measured. The experimental data for germanium and silicon, and for other semiconductors, show conclusively that radiation defects (even in the simplest, from the point of view of theory, case of electron bombardment) have complex spectra of shallow and deep energy levels in the forbidden band. The presence of several deep defect levels in a crystal with a wide forbidden band is explained qualitatively by extending the concept of the reduction in the electron binding energy in a medium with high permittivity to the case of multiple ionization (for example, to the case of interstitial atoms in germanium or silicon). Similarly, it is assumed that the unsaturated valence bonds in the region of a vacancy give rise to several electron capture levels. These ideas were first put forward by James and Lark-Horovitz [23], who suggested that an interstitial atom was a donor and a vacancy an acceptor.

As a rule, the generation of radiation defects in semiconductor crystals is accompanied by a change in the equilibrium den-

sity of carriers, because of carrier capture by defect levels or the
ionization of defects. In principle, information on the positions
and numbers of defect energy levels in the forbidden band may be
obtained by analyzing the change in the free carrier density per
unit integral particle flux Φ, i.e., the change in the value of
$\Delta n/\Phi$, or $\Delta p/\Phi$, for various positions of the Fermi level. Assuming that the number of defects is proportional to the flux, we
may write (for the special case when the electron density decreases in an n-type semiconductor):

$$\frac{-\Delta n + \Delta p}{\Phi} = \sum_m A_m f_m(E_F) - \sum_n D_n [1 - f_n(E_F)], \qquad (5.43)$$

where D and A are, respectively, the numbers of donor and acceptor levels produced per unit length of the particle path. The
function $f_i(E_F)$ is the probability that the i-th level is occupied by
an electron. It depends on the depth of the level and the degree of
its degeneracy. If one defect has several levels, $f_i(E_F)$ of a single level depends on the degree of population of the levels lying
at distances comparable with or less than kT from the level being
considered. However, when the levels are widely separated,
$f_i(E_F)$ represents an expression for the Fermi distribution and it
varies almost from 0 to 1 when the distance between the defect
levels and the Fermi level E_F is varied within 4 kT.

Therefore, if the levels are sufficiently wide apart, the
$(-\Delta n + \Delta p)/\Phi = \psi(E_F)$ curve should have discontinuities where the
level energies and E_F coincide [24]. To obtain the various positions of the Fermi level, it is, in principle, most convenient to
use a series of semiconductor samples containing different amounts
of the doping impurity (for example, a donor impurity). However,
the recently discovered interaction of chemical impurities with
defects in silicon complicates the interpretation of such experiments.

The temperature dependence of the Hall effect is frequently
used to determine the defect-level positions and the densities of
carriers captured or liberated by these levels. The use of the
Hall-effect method and the method of determining the relative
change in the carrier density as a function of the Fermi-level
position will be discussed in some detail, taking germanium and
silicon as examples.

Information on the defect energy levels may also be obtained by measuring the infrared absorption beyond the fundamental band edge. This optical method is particularly interesting because it gives information on the excited states of defects (centers) with deep levels. One of the difficulties of the optical method is the need to have high concentrations of the investigated centers (at least 10^{16} cm^{-3}). Moreover, it is sometimes necessary to cool the sample.

On the other hand, the investigation of the photoconductivity spectra which are associated with radiation defect levels, can sometimes be used to detect the energy levels of centers present in very small concentrations. It should be noted that in studies of the impurity photoconductivity careful control experiments are needed to distinguish the effects associated with centers in the interior of a sample from the phenomena due to photoionization of centers in the surface layer where the level spectrum may be different from that in the interior. The methods for investigating the impurity photoconductivity kinetics, developed recently by Ryvkin, Paritskii, and others [25], yield data on the cross sections of carrier capture by local levels in addition to data on level positions.

An important problem is the influence of the resultant defects on the velocity of recombination of nonequilibrium carriers. This problem is important in practical applications of semiconductors. Various experimental methods are used to determine the recombination parameters in studies of recombination centers produced by hard radiations. Examples of such methods will be described next. It is appropriate to note here that a change in the volume recombination velocity for nonequilibrium carriers is one of the most sensitive signs that defects are being produced in a semiconductor crystal, which may be convenient in "threshold" experiments on germanium and silicon single crystals.

§ 28. Radiation Defects in Silicon Single Crystals

Some of the important features of the energy-level spectra of radiation defects in silicon were observed in the early work carried out under the direction of Lark-Horovitz [4]. Since these experiments were carried out on crystal samples containing considerable amounts of chemical impurities, neutron irradiation in nuclear reactors and heavy particle bombardment were mainly used in order to produce high defect concentrations, necessary for the reliable recording of changes in the crystal properties.

The main qualitative result of their work (which, to a great extent, determined the technique of subsequent similar investigations) was the establishing of the following fact: irrespective of the sign of the electrical conductivity and of the type of chemical impurity (which governs the conductivity of silicon samples, before irradiation), structural radiation defects capture majority carriers. Consequently, the electrical conductivity of silicon decreases monotonically with increase of the integral radiation dose, approaching a value characteristic of intrinsic silicon   (about $2 \times 10^5 \ \Omega \cdot$ cm at 300°K).

A reduction in the electrical conductivity is primarily due to the fall in the carrier density; there is a simultaneous decrease in the mobility but it is a second-order effect.

A.   Threshold of Radiation-Defect Formation

All the known data on the threshold energies of defect formation in semiconductors, including silicon, have been obtained in experiments using monoenergetic electron beams of 0.1-1 MeV, because electrostatic generators and other types of electron accelerators which work in this range of energies are convenient sources of intense radiation beams. The disadvantage of electron bombardment is that on approaching the threshold energy the depth of the layer in which defects are formed decreases and tends to zero.

A change in the electrical conductivity is a sensitive "indicator" of the presence of radiation defects in a semiconductor. In the case of silicon, and even more so for germanium, the volume recombination velocity, which inreases on the appearance of the deep energy levels of defects, is an even more sensitive "indicator. "

No direct determinations of the threshold energy of defect formation in silicon have been carried out using changes in the electrical conductivity. According to Loferski and Rappaport [26], the minimum energy of fast electrons which would reduce the non-equilibrium carrier lifetime τ was 145 keV, corresponding to a threshold energy for atomic displacement of $E_d = 12.9$ eV. The method of determining τ in these experiments consisted of measuring the short-circuit current J_{pn} between the p- and n-type regions of a crystal with a p-n junction. The current J_{pn} was due to ionization by the fast-electron beam, which also produced radiation defects in the silicon. Under the experimental conditions, the value of J_{pn} was proportional to $\tau^{1/2}$. When the initial value of τ was sufficiently high, the appearance of 10^{10}-10^{12} cm^{-2} recombination centers reduced the value of τ considerably and this was easily detected from the drop in the current J_{pn}. However (see below), recombination centers produced in silicon by electron bombardment may be more complicated than Frenkel defects. Therefore, to solve finally the threshold-energy problem, it is necessary to compare the results of experiments carried out using various methods. The threshold-energy data based on changes in the electrical conductivity of electron-irradiated silicon were obtained by V. S. Vavilov, V. M. Patskevich, B. Ya. Yurkov, and P. Ya. Glazunov [27], who were able also to determine the dependence of the probability of generation of radiation defects on the mutual orientation of crystallographic axes of the sample and the direction of the incident beam. They used a monoenergetic electron beam of 500 keV, and silicon plates cut from one very uniform p-type single crystal (initial resistivity $\rho = 160 \ \Omega \cdot$ cm) in such a way that the beam was incident normally to the surface and parallel to the axes [111], [110], and [100]. The presence and the depth distribution of defects were found by comparing the potential distribution curves obtained before and after electron bombardment. * Control measurements of the Hall effect showed that the carrier mobility at room temperature changed so little that, assuming that defects which appeared at different depths were identical, one could conclude that the electrical conductivity was proportional to the defect concentration. Figure 76 shows the dependence of the resistivity ρ on the depth of penetration of the electron beam x for three different orientations of the single crystals. The maximum depth at which the resistivity rise was still observed was about 0.6 mm and it was independent of the crystal orientation; this can be explained in a natural way by the strong scattering of electrons in various directions as they penetrate the

* The sample was in the form of a rectangular slab with 1-μ thick electrodes at its ends. Passing a current along the sample and carrying out probe measurements, the authors were able to obtain sufficiently accurate data on the resistivity ρ and its dependence on the distance from the electron-bombarded surface.

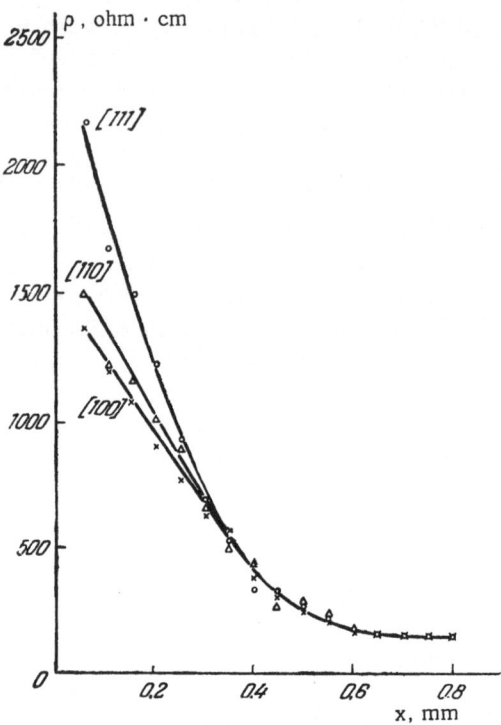

Fig. 76. Dependence of the resistivity ρ on the depth of electron-beam penetration x for three different orientations of silicon single crystals [27].

crystal. However, Fig. 76 shows also that near the surface the bombardment along the [111] axis caused a stronger rise in the resistivity than did the bombardment along the other two directions.

In order to estimate the value of E_d, the experimentally determined variation of the carrier density with depth below the surface [i.e., the dependence $-\Delta p = f(x)$] was compared with the theoretical curves plotted allowing for the electron scattering at threshold energies E_{min} equal to 280 keV* and 145 keV [26]. The curves were normalized at the point x = 0, which was permissible on the assumption that the population of the defect levels near the surface and in the interior was the same. Since, in fact, the Fermi-level

*In their earlier work, Loferski and Rappaport gave 280 keV, but this was replaced in [26] by 145 keV.

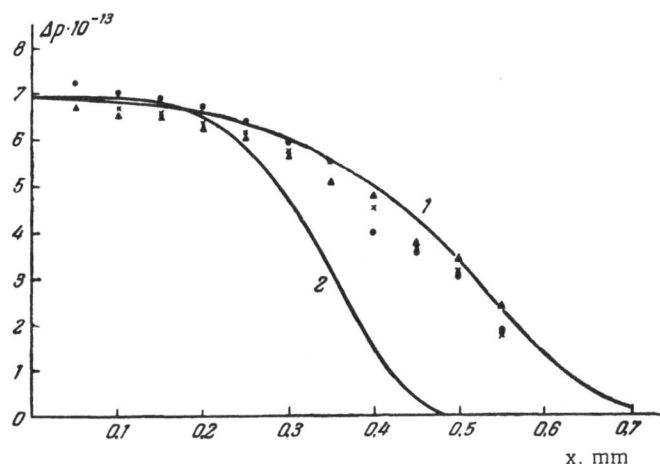

Fig. 77. Variation in the defect concentration with depth in silicon
single crystals (the three samples of Fig. 76) subjected to electron bombardment [27].

position varied with the number of defects, the latter assumption
has led to some errors. However, the data on the level spectra
of defects, which we shall consider below, showed that these errors are small.

The curves in Fig. 77 indicate that the theoretical plot for
145 keV (1) agrees much better with the experimental data than
does the defect-distribution curve for 280 keV (2).

On the other hand, Fig. 77 shows that, for diamond-type lattices, the theoretically predicted dependence on the orientation
appears near the crystal surface where the electron energy is
close to 500 keV, i.e., where $E > 3E_{min}$, and the majority of
scattering acts are accompanied by the transfer of momentum
to atoms at large angles, exceeding 60°, measured from the bombardment direction [111]. Therefore, we must assume that the
formation of stable radiation defects in silicon is accompanied by
atom displacement not along the [111] axis but along the axes
[010], [001], and [100].

B. Influence of Radiation Defects on the Equilibrium
Carrier Density

One of the qualitative conclusions drawn in the work just cited
was that silicon irradiated with large doses of heavy particles or

neutrons possessed shallow defect levels close to the conduction
and valence bands. Somewhat later, Wertheim [28] investigated
the effect of fast-electron bombardment on silicon, by measuring
the temperature dependence of the Hall effect, resistivity, minority-carrier and nonequilibrium-carrier levels. His experiments
showed that, apart from the shallow levels, deep levels were also
produced and they acted as carrier-capture centers. Similar data
on the energy levels of defects in silicon, including shallow levels,
were obtained by D. Hill [29] whose basic method was to determine the relative amounts of equilibrium carriers captured from
the bands by radiation defects. For this purpose, he used samples in which the initial carrier densities varied over a wide range.
Silicon was irradiated by electrons of energies ranging from 0.2
to 5.5 MeV. The bombardment was carried out in vacuum in a
cryostat in which the sample temperature could be varied from
10°K upward. Small integral electron doses were used and this
made it easier to interpret the results without taking into account
the Fermi-level displacement by the bombardment. The measurements gave the values of the Hall coefficient R, and the resistivity of irradiated silicon.

In the presence of carriers of one type only,

$$R = \pm \frac{sr}{nq}, \qquad (5.44)$$

where n is the carrier density. The numerical coefficients s and
r have values close to unity and are governed by the nonsphericity
of constant energy surfaces, and by the nature of carrier scattering. According to Abeles and Meiboom, s = 0.87 for n-type silicon; for p-type silicon, s = 1. The values of r are given by Johnson and Lark-Horovitz [31] for nondegenerate silicon, and by
Shipley and Johnson [32] for degenerate silicon.

The analysis of the experimental data is based on the dependence of the probability of filling defect levels on the Fermi-level
position E_F. We shall consider, by way of example, an n-type
semiconductor with chemical donors of one type; we shall assume
that the concentration of these donors is N_D and that practically
all of them are ionized at the test temperature. We shall further
assume that the bombardment produces defects   the density of
which is N_B, and that these defects have a single energy level in
the upper half of the forbidden band at a distance E_B from the

conduction band edge. Before the appearance of defects, the equilibrium electron density is n = N_D. After bombardment, the new
equilibrium density n' may be written as follows

$$n' = N_D - N_B \left[1 + \alpha \exp \frac{E_B - E_F}{kT} \right]^{-1}, \qquad (5.45)$$

where α is the ratio of the statistical weights of the unfilled to
filled state for the level considered, which is usually 2 [33]. Assuming that the number of defects is proportional to the integral
electron dose Φ, i.e., $N_B = A_B \Phi$, we can write down the expression for the relative change in n:

$$\frac{\Delta n}{\Phi} = \frac{n' - n}{\Phi} = A_B \left[1 + \alpha \exp \frac{E_B - E_F}{kT} \right]^{-1}, \qquad (5.46)$$

where the constant A_B, having the dimensions of cm^{-1}, represents
the "efficiency of defect generation in unit volume by unit flux."
As pointed out above, when several levels are generated, the right-
hand side of the above equation is the sum of a number of terms,
each of which represents a single level.

We shall return now to the case of a single defect level and
consider the expression for n'. It is evident that as the temperature is varied, the dependence of n' on T will be a stepped curve,
the height of the step representing the defect concentration N_B.
When the Fermi level coincides with the defect level, the carrier density will be $N_D - N_B (1 + \alpha)^{-1}$. If $N_B > N_D$, the slope of
the linear portion of the dependence $(\ln n) T^{3/2} = f(1/T)$ represents
the level energy E_B; in this case, the value of α is not used in
determining the level position.

Figures 78 and 79 give the results of Hill [29] for $-\Delta p / \Phi$
and $-\Delta n / \Phi$ for p- and n-type silicon samples subjected to bombardment with 4.5 MeV electrons at 0°C. * That the highest values
of $-\Delta p / \Phi$ and $-\Delta n / \Phi$ are observed on irradiation of degenerate
material, and that they decrease rapidly as the Fermi level moves
to the middle of the forbidden band, indicates that mainly shallow
levels are active. Had the effects been due to levels all lying at
the same depth, Eq. (5.46) could have been checked against the
experimental data.

• Hill used single crystals containing more than 10^{17} cm^{-3} oxygen atoms.

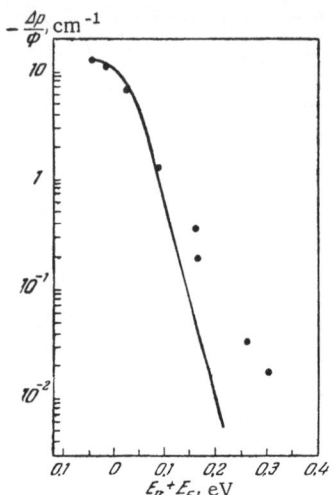

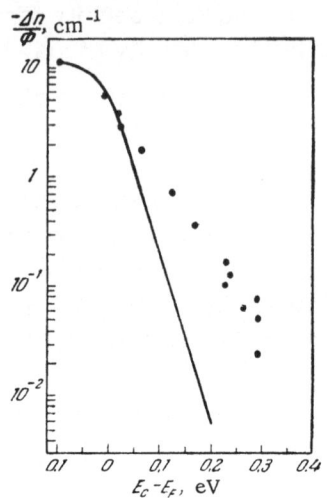

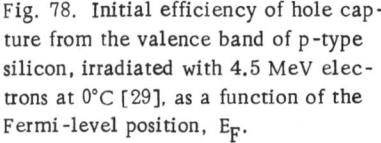

Fig. 78. Initial efficiency of hole capture from the valence band of p-type
silicon, irradiated with 4.5 MeV electrons at 0°C [29], as a function of the
Fermi-level position, E_F.

Fig. 79. Initial efficiency of electron
capture from the conduction band of
n-type silicon irradiated with 4.5 MeV
electrons at 0°C [29], as a function of
the Fermi-level position, E_F.

The continuous curves in Figs. 78 and 79 are plotted using
Eq. (5.46). These curves pass far below the experimental points
corresponding to the Fermi level, lying further than 0.1 eV from
the band edge, i.e., the reduction of the equilibrium carrier density is much greater than the value calculated on the assumption
that only shallow defect levels are produced. It is natural to assume that both deep and shallow levels are generated, but there
are fewer deep levels. The problem of the positions of deep levels
and of the nature of defects to which they belong will be considered
below.

Hill reported [29] that shallow defect levels in n-type silicon
were separated by 0.03 eV from the conduction band. The "efficiency" of generation of these levels was $A_B = 11$ cm^{-1}. In p-type
silicon, they were separated by 0.05 eV from the valence band and
their generation "efficiency" was 13 cm^{-1}.

Theoretical calculations of the probability of defect generation,
found using the formulas given at the beginning of the present chapter on the assumption that $E_d = 15$ eV, gave the rate of defect generation, which was about 7 cm^{-1} for 4.5 MeV electrons. For this

reason, it was natural to assign the shallow levels near the bands
to Frenkel-type defects.

The positions of deeper defect levels were determined by Hill
[29] and by several other workers [34] using the temperature dependence of the Hall coefficient and of the electrical conductivity.
At present, we can regard as reliably established the level positions in Fig. 80, where the data obtained by the temperature-de-
pendence method and by other methods are compared. The considerable differences in the rates of generation of various defect
levels do not agree with the simplest assumption that all levels
are due to Frenkel-type defects. It has recently been shown that
the rate of generation of individual defect levels is governed not
only by the energy and flux of the bombarding particles, but also
by the presence of chemical impurities, in particular, oxygen and
phosphorus. This has led to the suggestion that deep levels are
related to the interaction between structural point defects and impurities which form complexes with definite binding energies.
The problem of the nature of some of these complexes has been
solved by using paramagnetic (spin) electron resonance to study
silicon subjected to bombardment with high-energy particles.
Thus, for example, Bemski [35] investigated spin resonance in
n-type silicon irradiated with fast electrons and found absorption
lines close to 9000 Mc/s and 24,000 Mc/s for crystals grown by
pulling from the melt in quartz crucibles and containing about 10^{18}
cm^{-3} oxygen atoms. These lines were not detected in silicon crystals with less than 10^{17} cm^{-3} oxygen, which were prepared by
zone refining in vacuum. The absorption was independent of the
nature of the donor impurity (phosphorus or arsenic). A detailed
study of the spin-resonance spectrum, in particular its dependence on orientation and a comparison of the data obtained from the
information on defect energy levels, led Bemski to suggest a model for the resonating center. This model, shown schematically
in Fig. 81, is also based on the positions of oxygen atoms dissolved
in the silicon lattice (Chapt. I). According to Bemski, the vacancy
produced by the displacement of an atom from its site moves to
the stressed region near an oxygen atom occupying an interstitial
position. It is likely that the vacancy acts as an acceptor center,
capturing an unpaired electron, which leads to the resonance absorption of microwaves. The level at E_c -0.16 eV, the existence
of which has been confirmed by several independent methods, be-

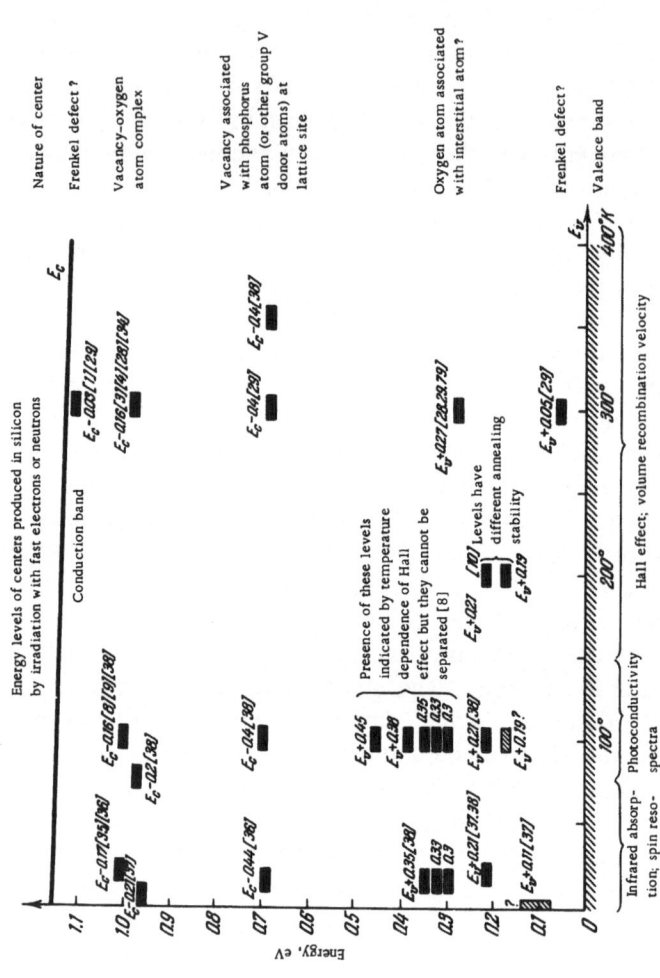

Fig. 80.   Energy-level scheme of radiation defects in silicon single crystals.

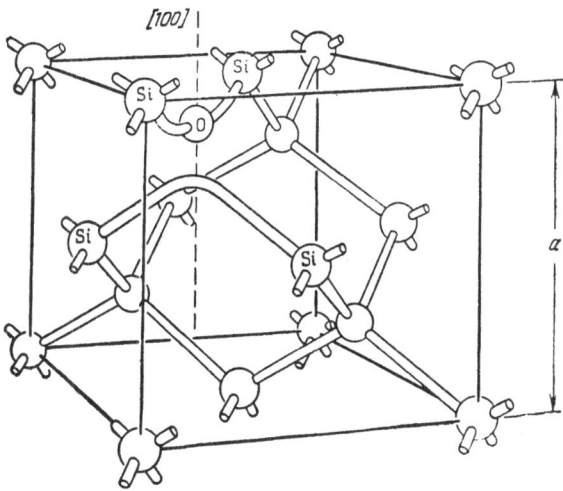

Fig. 81. Model of a complex center produced by electron
bombardment in silicon containing oxygen. A vacancy is
associated with an oxygen atom. In n-type silicon of low
resistivity, such a center captures electrons.

longs to this complex center. Watkins et al. [36] confirmed these
results and showed, moreover, that silicon crystals grown by
zone melting in vacuum without a crucible and irradiated with fast
electrons exhibited a maximum due to a center with an acceptor
level separated by about 0.43 eV from the conduction band. The
resonant absorption was observed only if the level was not filled
with an electron. It was found that this level belonged to a complex consisting of chemical-impurity atoms with a nuclear spin
of 1/2 and a simple defect, i.e., a vacancy or an interstitial silicon atom. The complexes probably consisted of mobile vacancies
and impurity atoms. The results of Watkins et al. [36] provide
an additional proof that the appearance of centers with deep levels
in irradiated silicon is the result of a primary act of energy transfer to an atom and the generation of a Frenkel defect followed by
the migration of point effects to distances considerably greater
than the interatomic spacing.

C. Infrared Absorption and Photoconductivity
Associated with Radiation Defects in Silicon

The investigation of infrared radiation and photoconductivity
spectra outside the fundamental absorption band has yielded inde-

pendent data on electron transitions and energy levels of radiation defects in silicon. The optical methods of investigation have been found convenient, mainly because an increase in the radiation dose, and a consequent increase in the defect concentration, is accompanied by a considerable decrease in the absorption by carriers, which is caused by carrier capture by defects, making it easier to detect selective absorption bands associated with defects.

Infrared absorption by silicon irradiated with high-energy particles was investigated by Fan and Ramdas [37] and by the staff of P. N. Lebedev Physics Institute of the USSR Academy of Sciences [38]. Apart from an increase in the transparency of silicon, the observed effects of radiation defects reduced to the following. In the range of wavelengths next to the fundamental absorption edge, the absorption increased. At longer wavelengths, selective absorption bands appeared with room-temperature maxima at 1.8, 3.3, 3.9-4.1, 5.5, 5.9-6.0, 20.5, 27, and 30.1 μ. Investigations of the photoconductivity spectra have shown that, under favorable conditions, it is possible to observe clear "steps," corresponding to photoionization thresholds, at 4.1, 3.8, 3.3, 2.8, 1.62, and 1.23 μ. To interpret these optical data, it is frequently necessary to know not only the type of conduction in a given crystal but also the Fermi-level position at the temperature of optical measurements. The experimental results will be considered briefly below.

n-Type silicon. Initially, the single crystals had a resistivity of 0.03-0.04 $\Omega \cdot$ cm and contained phosphorus. Irradiation was carried out in a reactor at a temperature not higher than 60°C. As in other work [39, 40], the irradiated silicon was found to have a wide absorption band next to the fundamental band and the maximum of this new band was 1.8 μ. The 1.8 μ band was not detected in n-type crystals whose Fermi level at the test temperature was above $E_c - 0.16$ eV. Weaker absorption bands had maxima at 3.3-3.7 and 5.5 μ. At temperatures close to 100°K and below, all the absorption bands became narrower. The 1.8 μ band did not exhibit any structure but the 3.7 μ band split into several components. According to Fan and Ramdas [37], similar splitting was also observed for n-type silicon irradiated with electrons, and at 4.2°K maxima appeared corresponding to 0.374, 0.359, and 0.343 eV. By analogy with the separations of the excited-state levels of group V donors in silicon, Fan and Ramdas [37] concluded that the observed system of bands might be due to electron excitation associated with a doubly charged center be-

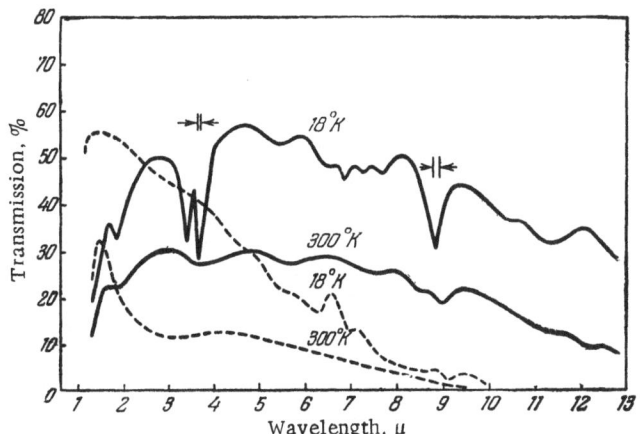

Fig. 82. Transmission spectra of n-type silicon before and after
neutron irradiation [38]. N-type silicon, ρ_0 = 0.04 Ω · cm;
dashed curves represent unirradiated, at a sample thickness d
= 0.05 cm; continuous curves represent silicon irradiated with a
a dose Φ = 6 x 10^{17} neutrons/cm^2, d = 0.05 cm.

cause the excited electron "orbits" were large compared with the
lattice constant.

The 5.5 μ band was obviously due to the photoionization of a
level separated by 0.2-0.16 eV from the conduction band edge
(Fig. 82).

The photoconductivity of n-type silicon irradiated with neutrons was also investigated by Fan and Ramdas [37], who did not
find the spectral features associated with the absorption band at
3.3-3.7 μ but noted a clear step at photon energies of 0.16-0.22 eV.

p-Type silicon. p-Type neutron-irradiated (as well as fast-
electron-irradiated [37]) silicon, containing boron as an acceptor
impurity, also exhibited the 1.8 μ band. The symmetrical pro-
file and other features of this band indicated that it was due to excitation and not to photoionization.

A band with a maximum at 3.9-4.1 μ was detected in p-type
silicon irradiated with neutrons when the Fermi level was not
more than 0.25 eV from the valence band.

A band with a maximum at 5.9-6.0 μ was also observed in
irradiated p-type crystals when the Fermi level at the test temperature was quite close to the valence band.

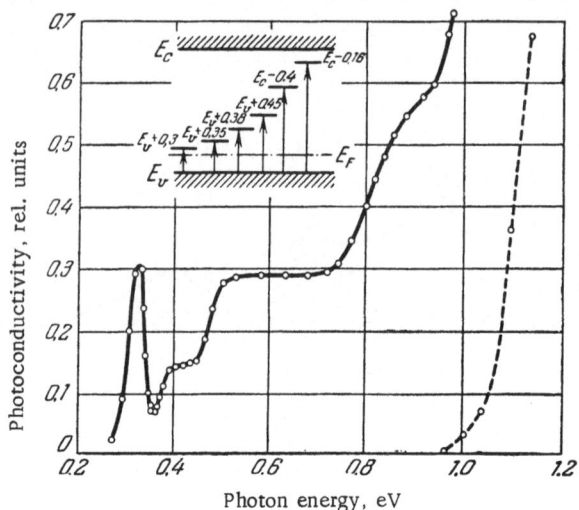

Fig. 83. Photoconductivity spectrum of a p-type silicon crystal, associated with radiation defects [38].

Finally, a band near 20.5 μ was found in crystals whose conductivity was near-intrinsic as a result of irradiation with a large integral neutron dose [37]. It has been suggested that the latter band represented one of the optical modes of the lattice vibrations of silicon, which was not excited in a crystal free of defects. The photon energy, 0.0605 eV, is close to the energy of Raman phonons in silicon, which was found by cold-neutron scattering measurement to be 0.063 ± 0.013 eV [41].

An investigation of the spectra and kinetics of the photoconductivity associated with radiation defects in p-type silicon was carried out at about 100°K, using smaller neutron or fast-electron doses than in the investigations of the infrared absorption. The integral fast-neutron dose did not exceed 10^{14}–10^{15} cm^{-2} [38]. Thick germanium or silicon filters were used to eliminate the scattered light in all measurements of the photoconductivity beyond the fundamental band edge. The photoconductivity spectra had clear steps, the positions of which (in [38] the "half-heights" of the steps were used) were employed to estimate the photoionization threshold energies of the defect levels.

In studies of the impurity photoconductivity, it is extremely important to separate the volume from surface effects, using sam-

ples of various thicknesses and comparing the photoconductivity
spectra with the calculated Fermi level positions. In p-type silicon irradiated with neutrons and electrons [38], the photoconductivity was a volume effect. According to the results obtained in
that work, the threshold photoionization energies of radiation defects were represented (close to 100°K) by the following level
positions:

$$
\begin{array}{l}
E_V + 0.3 \text{ eV } (4.1\ \mu) \\
E_V + 0.33 \text{ eV } (3.8\ \mu) \\
E_V + 0.38 \text{ eV } (3.3\ \mu) \\
E_V + 0.45 \text{ eV } (2.8\ \mu)
\end{array}
\left.\rule{0pt}{5em}\right\} \quad
\begin{array}{l}
\text{in the lower half of} \\
\text{the forbidden band}
\end{array}
$$

$$
\begin{array}{l}
E_c - 0.4 \text{ eV } (1.62\ \mu) \\
E_c - 0.16 \text{ eV } (1.23\ \mu)
\end{array}
\left.\rule{0pt}{2.5em}\right\} \quad
\begin{array}{l}
\text{in the upper half of} \\
\text{the forbidden band}
\end{array}
$$

Figure 83 shows the photoconductivity spectrum of neutron-irradiated p-type silicon crystals, recorded in the photon-energy
range down to 0.2 eV [42].

D. Influence of Radiation Defects on the Recombination of Nonequilibrium Carriers in Silicon

As pointed out in the foregoing the subsection, the generation of
radiation defects with deep energy levels increases the volume recombination velocity, i.e., it shortens the lifetime of nonequilibrium carrier pairs in silicon. This effect was used by Loferski
and Rappaport [26] to determine the threshold energy E_d necessary
for the formation of recombination centers. However, in this and
in other studies of the direct transformation of beta-ray energy
into electrical power by means of silicon crystals with p-n junctions, the investigators limited themselves to presenting data on
the dependence of the carrier lifetime on the integral particle
dose, without determining the level position or the capture cross
section. More detailed investigations carried out by G. Wertheim
[28], and by G. N. Galkin, N. S. Rytova, and V. S. Vavilov [34]
gave, in some cases, the physical parameters of the main recombination centers produced by irradiation of silicon with fast electrons, gamma rays, and neutrons, and related these parameters
to the data on the defects found by other methods.

In all this work, the recombination center parameters were determined by analyzing the temperature dependence of the non-equilibrium carrier lifetime using the Hall-Shockley-Read statistics for a single level. It was shown that if radiation defects were produced in silicon by fast electrons and gamma rays, the recombination proceeded via carrier capture either by the E_C − 0.16 eV level or by the E_V +0.27 eV level. The number of centers with these levels, found from control measurements of the Hall effect, increased proportionally to the total number of fast electrons incident on the crystal. A study of the carrier mobility in the irradiated crystals showed that when the Fermi level was above the E_C − 0.16 eV level, the structural defects responsible for the reduction in the mobility had single negative charges, and the number of scattering centers was equal to the number of electrons captured from the conduction band. When the Fermi level was below the E_V + 0.27 eV level, the defects responsible for the reduction in mobility had single positive charges. These results could be interpreted by assuming that the E_C − 0.16 eV level was of the acceptor type and the E_V + 0.27 eV level was of the donor type. However, the values of the electron and hole capture cross sections, obtained by Wertheim for these nonequilibrium carriers, indicated that a more complex model was required. The electron and hole capture cross sections of the E_V + 0.27 eV level ($\sigma_n = 9.5 \times 10^{-15}$ cm^2, $\sigma_p = 8 \times 10^{-13}$ cm^2) indicated an acceptor level. The electron and hole capture cross sections for the E_C − 0.16 eV level were equal ($\sigma_n \approx \sigma_p \approx 1.8 \times 10^{-15}$ cm^2) and were of the same order of magnitude as the cross sections of neutral atoms. In order to correlate the values of the cross sections with the data on the change of the mobility and equilibrium carrier density in electron-irradiated silicon, Wertheim assumed that the defects responsible for the change of the mobility were Frenkel defects. He assumed that not only acceptor centers but also donor centers could capture electrons if there were an acceptor level below a donor level. Similarly, holes could be captured both by a usual donor and by an acceptor level lying below donor levels. Then, according to Wertheim, the change in the hole mobility in p-type silicon might be due to the scattering of holes on the total charge of widely separated pairs (Frenkel defects), i.e., vacancies and interstitial atoms separated by not less than 50 A; the level E_V + 0.27 eV was assumed to belong to the acceptor center of the

Frenkel pair and, together with the level of the donor center lying near the conduction band, captured holes in p-type silicon. The scattering of electrons, which determined the change in their mobility in irradiated n-type silicon, was therefore due to the charge of closely spaced pairs (the distance between the centers being about 2.5 A) in which the donor center with the level at $E_c - 0.16$ eV, in combination with the acceptor level lying near the valence band, captured electrons from the conduction band. It follows that in each case a Frenkel defect had both a deep level and a shallow level, the latter close to the edge of one of the bands. Shallow levels, which were probably associated with Frenkel defects, were reported by D. Hill [29]; however, it was found that the concentration of centers with deep levels was considerably smaller than that of centers with shallow levels. Moreover, Hill's [29] and Wertheim's [28] data on the change in the mobility after irradiation were not in agreement. G. N. Galkin, N. S. Rytova, and V. S. Vavilov [34] investigated electron-irradiation of n-type silicon (resistivity of about $5 \, \Omega \cdot$ cm) containing a donor impurity (phosphorus) and dissolved oxygen, and found that the $E_c - 0.16$ eV level dominated recombination in the temperature range 200-400°K. Galkin et al. [34] used the method for determining carrier lifetime suggested by Kalashnikov and Penin [43], which was based on a determination of p-n junction parameters, governed by the carrier lifetime near the junction, as a function of the frequency of the applied voltage. The results of the experiments carried out on silicon crystals, irradiated with electrons (beta rays from an Sr^{90}-Y^{90} source with an average energy close to 1 MeV) and Co^{60} gamma rays were identical.

In contrast to the results of Wertheim, the cross sections for carrier capture by the $E_c - 0.16$ eV level were found to be unequal: $\sigma_p = 4 \times 10^{-14}$ cm^2, $\sigma_n = 1 \times 10^{-15}$ cm^2. Therefore, the recombination capture levels were of the acceptor type. The results obtained did not allow the use of the hypothesis of a "closely spaced" pair consisting of an interstitial atom and a vacancy. On the other hand, the conclusion about the acceptor nature of the $E_c - 0.16$ eV level was in agreement with the results cited earlier of a spin-resonance study of irradiated silicon [35, 36], according to which this capture level was due to a complex consisting of a vacancy and an oxygen atom, and not due to an isolated vacancy or an interstitial atom.

TABLE 10. Energy Levels of Radiation Defects in Silicon*

Level position, eV	Type of center	Nature of radiation	Rate of introduction of levels, cm^{-1}	Experimental method	Capture cross sections at quoted temperatures	References†
$E_c - 0.03$	□	e (4.5 MeV)	11	C		[1] D. Hill, Phys. Rev. 114, 1414 (1959)
$E_c - 0.025$	□	d (9.6 MeV)	670	C		[2] T. Longo and K. Lark-Horovitz, Bull. Am. Phys. Soc., Ser. II, 2, 157 (1957)
	□	γ (Co⁶⁰)	~0.001	H		[3] E. Sonder and L. C. Templeton, J. Appl. Phys. 31, 1279 (1960)
	⊕	e (0.7 MeV)	0.18	H, R	$\sigma_{n,p} \approx 2 \times 10^{-15}$ cm² (300°K)	[4] G. Wertheim, Phys. Rev. 105, 1730 (1957); 110, 1272 (1958)
$E_c - 0.16\,(0.17)$	□	e (1.0 MeV)	—	P		[5] V. S. Vavilov and A. F. Plotnikov, FTT, 3, 2455 (1961)
	□	e (4.5 MeV)	0.5	H		[1]
	○	e (0.5 MeV)	—	S		[6] G. Bernski, J. Appl. Phys. 30, 1195 (1959)
	○	e (1.5 MeV)	—	S		[7] G. Watkins, J. Corbett, and R. Walker, J. Appl. Phys. 30, 1198 (1959)
	○	n (reactor)	—	H		[8] C. Klein, J. Appl. Phys. 30, 1222 (1959)
	○	e (Sr⁹⁰–Y⁹⁰)	—	H, R	$\sigma_p = 4 \times 10^{-14}$ $\sigma_n = 1 \times 10^{-15}$ (300°K)	[9] G. N. Galkin, N. S. Rytova, and V. S. Vavilov, FTT, 2, 9 (1960).

TABLE 10 (Continued)

Level position, eV	Type of center	Nature of radiation	Rate of introduction of levels, cm^{-1}	Experimental method	Capture cross sections at quoted temperatures	References†
$E_c - 0.21$	①	Co^{60}		R	$\sigma_p = 2.5 \times 10^{-14}$	[10] G. N. Galkin, FTT, 3, 630 (1961)
		n, e		I	$\sigma_n = 0.8 \times 10^{-15}$ (300°K)	[11] H. Y. Fan and A. Ramdas, J. Appl. Phys. 30, 1195 (1959)
						[1]
$E_c - 0.40$	☐	e (4.5 MeV)	0.05	C, H		[7]
		e (0.7 MeV)	0.004			[5]
Levels close to middle of forbidden band		e (1.0 MeV)		H, S		[12] G. Wertheim, J. Appl. Phys. 30, 1166 (1959)
		e (1.0 MeV)		P		
		n (fission of uranium)		R		
$E_v + 0.45$		e (1.0 MeV)		H, R	$\sigma_p \approx 10^{-17}$ (100°K)	[5]; [13] A. F. Plotnikov, V. S. Vavilov, and L. S. Smirnov, FTT, 3, 3253 (1961)
		n (reactor)				
$E_v + 0.38$		e (1.0 MeV)		P	$\sigma_p \approx 5 \times 10^{-17}$ (100°K)	[5]; [13]
$E_v + 0.33$		n (reactor)		R		[14] G. Wertheim, J. Appl. Phys. 30, 1166 (1959)
Levels close to middle of forbidden band		e (1.0 MeV)				
		n (fission of uranium)				
$E_v + 0.45$		e (1.0 MeV)		H, P	$\sigma_p \approx 10^{-17}$ (100°K)	[5]; [15] A. F. Plotnikov, V. S. Vavilov, and L. S. Smirnov, FTT, 3, 3253 (1961)
		n (reactor)		H, P		

TABLE 10 (Continued)

Level position, eV	Type of center	Nature of radiation	Rate of introduction of levels, cm⁻¹	Experimental method	Capture cross sections at quoted temperatures	References†
$E_V + 0.38$		e (1.0 MeV)		P	$\sigma_p \approx 5 \times 10^{-17}$ (100°K)	[5]; [15]
$E_V + 0.33$		n (reactor)		P		
	⊞	e (1.0 MeV)		P		[13]
	⊖	n (reactor)		P	$\sigma_p = 8 \times 10^{-13}$	[8]
$E_V + 0.31$		n (reactor)	0.35	H	$\sigma_n = 9.5 \times 10^{-15}$ (300°K)	[4]; [14]
$E_V + 0.30$		e (0.7 MeV)	0.005	H, R		
		e (4.5 MeV)	0.3	C, H		
$E_V + 0.27$	⊞	e (0.7 MeV)	0.05	H, P		[1]
		n (reactor)		H, P, R	$\sigma_p = 3 \times 10^{-14}$ (100°K)	[5]; [13]
		e (1.0 MeV)		H, P		[5]; [4]
$E_V + 0.21$	⊞⊞	e (1.0 MeV)	0.03	H, P		[16] V. S. Vavilov, V. M. Malovetskaya, G. N. Galkin, and A. F. Plotnikov, FTT, 4, 1969 (1962)
$E_V + 0.19$		e (1.0 MeV)	0.007	H, P		
$E_V + 0.16$	⊞⊞⊞⊞	n (reactor)	0.35	H		[8]
$E_V + 0.05$		e (4.5 MeV)	13	C		[1]
$E_V + 0.06$		d (9.6 MeV)	750	C		[2]
		n (reactor)	< 0.65	H		[8]

•The following notation is used in the table: ⊟ – electron capture; ⊞ – hole capture; ⊖ – acceptor; ⊕ – donor. Nature of radiation: γ – gamma rays; e – electrons; d – deuterons; n – neutrons. Experimental methods: C – carrier capture; H – temperature dependence of the Hall effect; R – temperature dependence of the volume recombination velocity for nonequilibrium carriers; I – infrared absorption and Hall effect; P – photoconductivity spectra; S – electron-spin resonance and Hall effect.
†The numbering of these references is specific to this table and has no relationship to the references in the text.

Investigation of the recombination of nonequilibrium carriers in silicon irradiated with fast neutrons in a reactor [44], showed that the dominant levels in these crystals lay close to the middle of the forbidden band; this was in sharp contrast to the data on the irradiation with electrons and gamma rays. According to Wertheim [44] (whose conclusions were based mainly on the strong dependence of the lifetime on the injection level), the depths of the levels were different. A detailed analysis of the recombination capture was not given by Wertheim. Data was also available on the change in silicon transistor parameters after irradiation in a reactor [45], but the positions and parameters of recombination levels were not determined. Only the dependence of the lifetime in the transitor base, τ, on the integral neutron dose Φ was given: for n-type silicon this empirical dependence has the form $\tau = [(3 \times 10^6)\Phi^{-1}]$ sec, and $\tau = [(2 \times 10^6)\Phi^{-1}]$ sec for p-type silicon. These results gave values of τ which were about an order of magnitude greater than the data of Wertheim. Unfortunately, in many of the experiments carried out in reactors, without any special precautions, the heating of the crystals and the resultant annealing of the radiation effects were not allowed for.

It is too early to give a detailed quantitative description of the properties of radiation defects in silicon using a single model. The main reason for the difficulties is that the actual system of structural defects is determined not only by the primary process of energy and momentum transfer to silicon atoms but also by the interaction of vacancies and interstitial atoms with impurities. However, the data on the level scheme of radiation defects (Fig. 80) show satisfactory agreement between independent methods. An interesting point is the good agreement between the data on a level produced by neutron and fast-electron irradiations, which indicates that defects are far apart in those groups which are formed by the action of neutrons. The only worker whose results are not in agreement with this conclusion is Wertheim [44]. Table 10 lists, as a supplement to Fig. 80, the data of various authors on the positions of radiation-defect levels in the forbidden band.

§ 29. Radiation Defects in Germanium
Single Crystals

In spite of the fact that studies of radiation defects in germanium crystals have attracted much attention, their physical nature, stability, and energy-level positions are not yet completely established. In contrast to experiments on silicon, where information has been obtained by optical and electron-resonance methods, studies of germanium have been concerned mainly with changes in the carrier density and recombination through defect levels. Reproducible results have been obtained initially for n-type crystals; the changes in the physical properties induced by the bombardment of p-type crystals have usually been found to be unstable since defects are annealed intensively at temperatures as low as 300°K and sometimes even below this temperature. The first facts established in irradiation experiments were the reduction in the equilibrium electron density in n-type germanium crystals and their conversion to p-type conduction by sufficiently large integral doses. This has been used to produce p-n junctions in germanium crystals.

A. Dependence of the Probability of Radiation-Defect Formation on the Energy Transferred to Atoms during Electron Bombardment

The bombardment of germanium with fast electrons, for the purpose of detecting the "threshold" of radiation-defect formation, was employed first in 1950 [4]. The appearance of radiation defects was deduced from the change in the equilibrium electrical conductivity of relatively thick (0.5 mm) n-type germanium plates, subjected to bombardment with monoenergetic electrons at 78°K. The threshold energy was E_{min} = 630 keV (E_d = 30 eV). Later, Loferski and Rappaport [80] carried out experiments on the irradiation of n-type germanium with electrons of energies ranging from 0.3 to 1 MeV. The presence of radiation defects was deduced from the reduction in the nonequilibrium carrier lifetime. Under the experimental conditions employed by Loferski and Rappaport, the short-circuit current in the external circuit of a crystal with a p-n junction was proportional to $\tau^{1/2}$ during irradiation. In good-quality single crystals of germanium, the nonequilibrium carrier lifetime was governed by recombination centers whose concentration did not exceed 10^{10}-10^{12} cm^{-3}. The appearance of radiation defects in a concentration of 10^{12} cm^{-3}, which had recombination capture levels, reduced the lifetime to about one half, when the capture cross sections of the radiation defects and the capture cross sections of the initially present centers were similar; at the same time, the introduction of radiation defects altered the equilibrium carrier density by only 10^{12} cm^{-3}. Initially, this method gave a value of E_{min} equal to 500 keV, but later, in 1958 [26], the same workers reported a new value, which was 360 keV (i.e., E_d = 14.5 eV). In the work carried out in 1955 at the P. N. Lebedev Physics Institute of the USSR Academy of Sciences [46], an attempt was made to determine the threshold of Frenkel-defect formation in n-type germanium and to investigate the change in the relative concentration of defects at incident electron energies higher than the threshold value. The dependence of the defect-formation cross section on the fast-electron energy was plotted (Fig. 67) from a series of measurements of the resistivity of thin (50 μ) single-crystal plates of n-type germanium, subjected to bombardment with monoenergetic electrons from an accelerator. It was assumed that each defect captured one electron from the conduction band at 300°K (in test samples with ρ < 25 $\Omega \cdot$ cm). It was found that the experimental points lay considerably below the theoretical curve, which was possibly due to the low capture level population (the level positions were not determined by independent methods) or due to special features of the intrinsic process of defect formation, which were not allowed for in the theory. These measurements gave a threshold energy of 0.5 MeV; later, Brown and Augstyniak [12] and, independently, Smirnov and Glazunov [13], carried out careful measurements and concluded that the threshold energy was close to 360-380 keV. Brown and Augustyniak established that near the threshold energy (in the 370-420 keV range) the probability of defect formation by electrons incident along the [111] axis was higher than for other orientations of the sample. The difference between the probabilities for the directions [111] and [110] reached 40% at E = 420 keV. However, in contrast to what one would expect from the theory, on the approach of the energy E to the "threshold" value, the effect of the orientation decreased. Moreover, small changes in the carrier density also occurred at incident electron energies considerably lower than the "threshold" value.

The observed effect was a volume one but up to now it has not been established whether it is related to weakly bound atoms near dislocations or to impurities.

Experiments described in [12, 46] (and in the Soviet paper cited above) established that the concentrations of defect carrier-capture centers were, at incident electron energies above the threshold value, considerably lower than the concentrations calculated using the simple theory, which predicted the existence of a threshold E_{min} and assumed a unique probability of the formation of a stable defect when $E_A > E_d$. Brown and Augustyniak gave a semi-empirical curve which described well the experimental data and was plotted on the assumption that a defect was always stable at $E_A > 60$ eV and that the probability of "stabilization" of a defect increased smoothly in the range 15 eV < E < 60 eV [12]. A comparison of carrier density measurements after electron bombardment at various temperatures showed that, in general, below 300°K the number of defects decreased on cooling, other conditions being equal. This effect, completely analogous to that observed for silicon, was ascribed by Wertheim [45] to the instability of initially formed closely spaced Frenkel pairs and to the fact that the activation energy necessary for the destruction of a defect is somewhat lower than the minimum energy needed for the separation of a vacancy and an interstitial atom.

B. Energy Levels of Radiation Defects in Germanium and the Influence of Defects on the Recombination of Nonequilibrium Carriers

The methods for determining the defect-level positions, which are used in studies of irradiated germanium, are essentially analogous to those employed in the case of silicon.

Comparing the results of fast-neutron irradiations with those obtained using electrons and gamma rays, it is seen that the levels systems are different. The most likely explanation of this is the interaction of defects in closely knit groups or aggregates in the former case, which does not occur in crystals irradiated with electrons and gamma rays.

The system of energy levels produced by fast-neutron bombardment of germanium is shown in Fig. 84 [47]. *

In the well-known review of Lark-Horovitz and Fan [24], a more complicated system of levels is given, some of which were determined from photoconductivity data. However, Spear [48] later showed that some of these levels were not due to radiation defects in the interior but due to impurity centers at the surface of the germanium. We shall restrict ourselves to deep levels which are known to be associated with radiation defects in the interior of germanium single crystals.

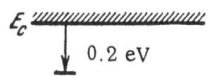

Fig. 84. System of energy
levels of defects produced
in germanium by fast-
neutron bombardment [47].

Cleland, Crawford, and Pigg [47]
also investigated the recombination of
nonequilibrium carriers for various defect concentrations and temperatures.
According to their interpretation, there
is a recombination level at 0.23 eV below
the conduction band. Their measurements were carried out on homogeneous
germanium slabs into which carriers
were injected from a surface-barrier
contact. The hole-capture cross section
of this recombination level was found to
be close to 3×10^{-15} cm^2.

Similar experiments were carried out by Messenger and
Spratt [49] on germanium p-n-p transistors irradiated in a reactor; they found that the hole-capture cross section was close to
10^{-15} cm^2 and the electron-capture cross section was almost
4×10^{-15} cm^2. Messenger and Spratt also concluded that the level
considered lay in the upper half of the forbidden band, 0.23 eV
from the conduction band. The small difference between the hole-
and electron-capture cross sections suggested that the $E_c - 0.23$
eV level was not involved in transitions between singly and doubly
charged states of a defect center.

V. S. Vavilov, L. S. Smirnov, A. V. Spitsyn, and M. V.
Chukichev [50] investigated recombination in n-type germanium
irradiated with monoenergetic 14 MeV neutrons in the absence of
gamma-ray background. According to their data (see also [51]),
the recombination level of radiation-induced centers lies in the
upper half of the forbidden band and the maximum possible hole-
capture cross section amounts to 10^{-15} cm^2.

Curtis and Cleland [52] also investigated the irradiation of germanium with monoenergetic 14 MeV neutrons; they concluded that
the dominant recombination level lay close to the middle of the
forbidden band and the center was of the acceptor type. This result is at variance with the data obtained by other workers. According to Curtis, Cleland, and Crawford [53], the $E_c - 0.23$ eV
level governs the principal recombination transitions in p-type
germanium as well, and it belongs to a defect with a double level in
the lower half of the forbidden band. Consequently, the displacement of the Fermi level so that it passes through the position of

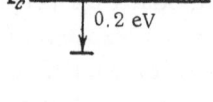

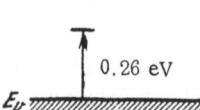

Fig. 85. System of deep energy levels of defects produced in germanium by irradiation with Co⁶⁰ gamma rays (1.17 and 1.138 MeV) [56].

the lower level may strongly affect the recombination velocity, as in other cases of multicharged centers (for example, nickel in germanium) [54]. If the Fermi level lies several kT below the first (lower) level of the defect, the $E_c - 0.23$ eV level is absent and, therefore, the lifetime in p-type crystals at low temperatures is considerably higher than the value calculated by means of the usual formulas on the assumption that the $E_c - 0.23$ eV level is always present.

Assuming that the defect is a multicharged center, the data on the capture cross sections of neutron-irradiated n- and p-type germanium can be made to agree [45].

Hasiguti and others [55] examined the influence of deuteron bombardment on the carrier lifetime in germanium but gave no information on the level positions, therefore it was difficult to compare their results with the data of other workers.

The level positions deduced from the Hall effect and from recombination experiments agree satisfactorily. On the other hand, the reported efficiency or "velocity" of defect generation by neutron bombardment differs very markedly from one author to another, which is probably due to the low precision of integral dose measurements and to the probable errors in allowing for heating during irradiation in a reactor.

We shall consider below the consequences of electron bombardment and gamma irradiation of germanium. Figure 85 shows the deep energy levels of defects produced in germanium by gamma rays [56].

Obviously, electron bombardment should produce the same system of levels.

According to the experiments of Curtis et al. [53], the dominant recombination level is an upper level at $E_c - 0.20$ eV. The position of this level is almost identical with the level of defects generated by neutron bombardment, but this time the hole-capture cross section is only 4×10^{-16} cm², i.e., it is one order of magnitude smaller than in the case of neutron bombardment. Investigations of recombination in irradiated p-type germanium forced

Curtis et al. [53] to assume that the $E_c - 0.20$ eV level was the upper level of a defect which possessed another deeper level. The difference between the cross-section values for the two types of bombardment may be due to the interaction of individual defects in defect aggregates (in the case of neutron bombardment); this interaction should not occur in the case of electron or gamma-ray irradiation. According to V. S. Vavilov and L. S. Smirnov [57], the hole-capture cross section is close to 10^{-16} cm^2. Later, careful measurements of the temperature dependence of the lifetime in electron-irradiated n-type germanium led L. G. Sharendo and L. S. Smirnov [58] to the conclusion that the dominant process was the recombination capture by the $E_v + 0.26$ eV level; according to their data, the hole-capture cross section was close to 6×10^{-15} cm^2, indicating that the center was of the acceptor type. Baruch [59] irradiated germanium with 2 MeV electrons and concluded that the recombination level lay near $E_c - 0.18$ eV. Ryvkin, Khansevarov, and Yaroshetskii [60] investigated the impurity photoconductivity associated with radiation defects in n-type germanium irradiated with gamma rays, and established that the photoionization energies of the two defect levels corresponded to their positions near $E_c - 0.2$ eV and $E_v + 0.26$ eV.

Data on the energies of electron transitions via radiation-defect levels were obtained also by investigating recombination radiation spectra of n-type germanium irradiated with electrons [61].

The energies of photons emitted in the impurity band of irradiated germanium corresponded to level positions 0.20-0.25 eV from the nearest band. The positions of the principal levels of radiation defects in n-type germanium have thus been reliably established but the problem of the nature of those centers to which these levels belong still requires clarification.

At sufficiently low temperatures, beginning with 230°K, holes are captured by traps in n-type germanium containing radiation defects. According to Shulman [62], hole-capture levels lie near $E_v + 0.28$ eV; the capture cross section at 200°K is close to 6×10^{-16} cm^2 and varies exponentially with temperature (the activation energy is $\Delta E \approx 0.05$ eV).

§ 30.   Radiation Defects in Semiconducting Compounds

In view of the wide range of possible practical applications of semiconducting compounds, the problem of the effect of hard radiations on these compounds is of major importance. A theoretical analysis of possible point defects in compounds is considerably more complex than in elements. For example, in a binary compound with the zinc-blende structure, the displacement or substitution of atoms may give rise to eight types of point defect: two types of vacancy, four types of interstitial atom (an interstitial atom may be surrounded by similar or dissimilar atoms), and two types of substitution defect.

The difficulties in analyzing these defects force us to approach the problem phenomenologically, using the results for germanium and silicon reported above. Having determined the change in the equilibrium density of carriers, which are captured from one or the other band by radiation-defect levels, we can deduce the defect concentration and capture level positions. In the initial estimates, it is very convenient to use degenerate or almost degenerate samples. For example, if the carrier density in an n-type material decreases as a result of irradiation, it follows that radiation defects have levels which are mainly of the acceptor type. Similarly, a reduction in the hole density in a p-type material indicates the appearance of donor centers. The experimental data indicate that the effect of hard radiations in all the semiconductors studied so far is to produce both donor and acceptor centers. As the number of radiation defects increases to values considerably greater than the initial concentrations of chemical impurities, the equilibrium carrier density and the associated Fermi level position approach limiting values (saturation). These limiting values are determined by the system of energy levels of the radiation defects.

The change in the electrical conductivity of indium antimonide due to bombardment with fast neutrons was investigated by Cleland and Crawford [63]. Figure 86 shows their results for n- and p-type InSb. The curves in Fig. 86 indicate clearly that initially the electrical conductivity σ decreased steeply (mainly due to a reduction in the carrier density). The irradiation of n-type crystals was accompanied by a monotonic reduction in the electrical conduc-

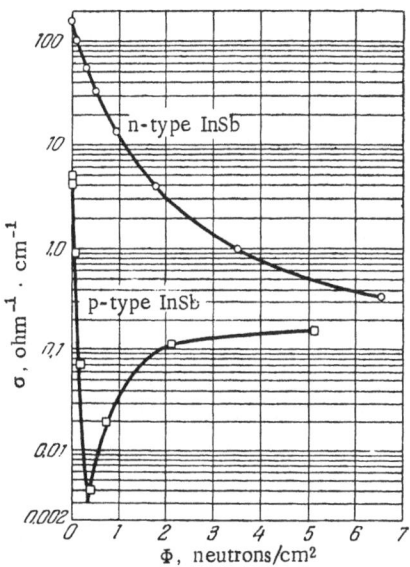

Fig. 86. Variation of the electrical conductivity of indium antimonide during neutron bombardment in a reactor at room temperature;
measurements were carried out at 77°K [63].

tivity, but in p-type crystals the irradiation changed the type of
conduction; at high doses, both the p- and n-type curves tended to
a limiting value of σ lying near $1 \, \Omega^{-1} \cdot \mathrm{cm}^{-1}$ at 77°K. An interpretation of the data on the irradiation of InSb with neutrons is
complicated by the appearance of donor centers as a result of nuclear reactions; however, there are reasons for believing that
even in the absence of these centers a conversion of p-type into
n-type crystals should occur although the electrical conductivity
at high defect concentrations should be somewhat lower than the
observed value.

It was found that defect energy-level systems appearing in
InSb after bombardment with electrons differed greatly, depending
on the temperature at which the bombardment was carried out.
It is very likely that the same might occur in the case of neutron
irradiation because, as far as is known, no experiments on the
irradiation of InSb with neutrons at low temperatures have yet
been carried out.

TABLE 11

Substance	n-Type samples, $-\Delta n/\Delta \Phi$, cm^{-1}	p-Type samples, $-\Delta p/\Delta \Phi$, cm^{-1}	Conduction type after a large dose	Average atomic mass	Carrier density before irradiation, cm^{-3}
	(at beginning of irradiation)				
SiC	2	—	—	20	10^{18}
SiC		5	intrinsic	28	10^{16}
GaAs	4	—	intrinsic	72	10^{16}
InP	5	—	—	73	10^{16}
Ge	3	3		73	10^{16}
AlSb	6	4.5	—	75	10^{15} - 10^{17}
GaSb	3	0.4		86	10^{17}
CdTe	—	1		120	10^{16}

Data on the effect of reactor neutron irradiation of n-type GaAs crystals [64] indicated that in this semiconductor the carrier density decreased steeply at high defect concentrations, approaching a value characteristic of intrinsic conduction.

A similar effect was noted in experiments on the irradiation of AlSb crystals when the surface layer of irradiated crystals was removed by, say, sand-blasting.

Approximate values of the relative changes in the carrier density in some of the semiconducting compounds irradiated with fast neutrons near room temperature are listed in Table 11 [64].

It is worth noting that the values of $-\Delta n/\Delta \Phi$ and $-\Delta p/\Delta \Phi$ are of the same order of magnitude in different compounds and that, moreover, the results for monoatomic (elemental) semiconductors do not differ greatly from those for compounds. This is an argument in favor of the conclusion that the number of collisions ending with a substitution is not much greater than the number of collisions which produce the usual defects.

According to Cleland and Crawford [65], the electron density in the conduction band of InAs increases (in contrast to many other semiconducting compounds) with increase of the dose both of neutrons and electrons. This occurs also in the irradiation of degenerate n-type samples. The results given here may be useful in further studies and in practical applications, but no information is available on the energy levels of radiation defects in in-

dium arsenide. Such information is known at present only for indium antimonide irradiated with 4.5 MeV electrons [66]. The
data presented below [66] refer to radiation defects which anneal
at temperatures as low as 300°K.

The temperature dependence of the electron density in n-type
crystals cannot be interpreted by means of a system of discrete
levels. The most likely explanation is the existence of levels of
various depths grouped in the region from $E_c - 0.04$ eV to E_c
$- 0.1$ eV. It should be noted that heavy doses of electrons convert
n-type InSb into p-type material, i.e., the final effect is opposite
to that observed under neutron bombardment at $T \approx 300°K$. The
nature of the centers produced by the irradiation of InSb and the
influence of impurities on these centers have not yet been studied.
Some data on the annealing of radiation defects in InSb and its
compounds are presented in the work of Aukerman and Lark-
Horovitz [66]. However, these results should be considered as
preliminary.

§ 31.   Annealing of Radiation Defects
in Semiconductors

A crystal in which radiation defects are present is unstable
and, in general, with time it will tend to return to its stable state
at any temperature other than 0°K; this means that the number of
defects will decrease. Experimental studies of the annealing of
radiation defects in crystals give information on the motion of
vacancies and interstitial atoms and on the interaction of these
simplest point defects with other imperfections – in particular,
with chemical impurities and dislocations. Detailed studies of
the reestablishment of the structure disturbed by the irradiation
of germanium and silicon have shown that the annealing processes
depend strongly on the nature and concentration of chemical impurities in the crystals. Thus, the initial theoretical interpretations, which assumed that only structural point defects exist and
interact in a crystal, were very approximate [67].

We shall consider qualitatively the processes of defect annealing in germanium and silicon crystals. Considering annealing as a diffusion-limited reaction, the simplest case is the annealing of defects produced by irradiation with fast electrons or
with gamma rays, when the Frenkel defects generated initially
can be considered to be uniformly distributed throughout the crystals.

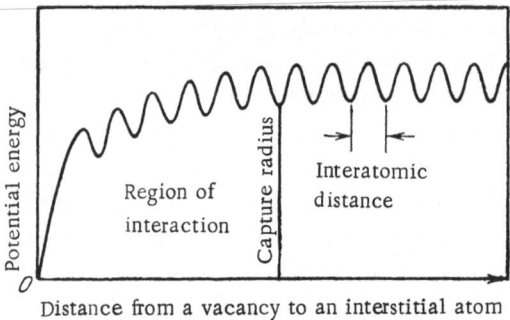

Fig. 87.   The simplest energy scheme used to interpret annealing of radiation defects.

The curve in Fig. 87 shows the dependence of the potential energy of a Frenkel defect on the distance between the interstitial
atom and the vacancy.

At very short distances between the vacancy and the interstitial atom the energy rises steeply, which corresponds to the
work necessary for the displacement of an atom from its site; the
first shallow minimum represents the transfer of an atom into one
of the nearest "natural" interstitial positions.   The subsequent
shallow minima of the curve in Fig. 87 represent interstices which
are separated by various distances from the original site of the
atom.   At very large distances from the original site, the depths
of these minima should be the same.   However, we must assume
that near the vacancy the displaced atom is acted upon by a force,
due to the distortion of the lattice, directed toward the vacancy.
The action of this force reduces the energies of the minima close
to the vacancy.   As pointed out at the beginning of the present
chapter, with reference to Kohn's detailed theory for a diamond
lattice, the nearest interstices along the [111] axis may represent
very shallow potential wells and are probably unstable.   No detailed reliable quantitative data are available as yet on the extent
of the region of lattice distortion and strong interaction between
a vacancy and an atom, but indirect estimates for diamond-type
lattices show that it cannot exceed several interatomic spacings
[67].

As shown earlier, electrons of energies close to 1 MeV may
transfer to germanium or silicon nuclei energies of the order of
100 eV.   Collisions accompanied by maximum energy transfer are

relatively rare, and in the majority of cases atoms are knocked
out of their states as a result of energy transfer, which is only
slightly greater than the threshold value. The natural scatter of
the values of the momentum transferred to the displaced atoms
and of the directions of motion of these atoms lead to the occupation of various interstitial states shown schematically by the curve
of Fig. 87. In close-packed crystals, such as copper, the separation between the Frenkel-pair constituents may increase considerably as a result of "focused" collisions along certain crystallographic axes. However, in diamond-type lattices such processes are not very likely.

Thus, the initial result of electron bombardment should be
the appearance of relatively closely spaced Frenkel pairs. During
the subsequent motion of Frenkel defects, the probability of the
constituents coming together is higher than the probability of a
further increase in the separation. The activation energy representing the motion of defects should be less for closely spaced
pairs than for those separated by larger distances. Investigation of the annealing of radiation defects in metals (copper) has
made it possible to study separately the process of healing of closely spaced Frenkel defects [68]. It is probable that the healing
processes observed in germanium crystals initially bombarded
at low temperatures are also due to the disappearance of closely
spaced Frenkel pairs [69].

Defects separated by distances greater than the radius of action of attractive forces, which are due to lattice distortion or to
electrostatic forces, should move at random in a crystal due to
thermal activation. If the interstitial atom is more mobile than
the vacancy, it may, jumping from one interstice to another, ac-
cidentally find itself within the sphere of capture of a vacancy;
this will increase the jump frequency and the pair will be annihilated.

The motion of all the interstitial atoms relative to all the
vacancies (or the relative motion of the atoms and the vacancies)
should lead to a second-order healing process, i.e., to a hyper-
bolic dependence on time if the concentrations of the recombining
particles are equal. A theory of processes of this kind has been
developed by Antonov-Romanovskii [70], and Fletcher and Brown
[11], and generalized by Waite [71].

Naturally, in those cases when a Frenkel pair recombines
only after point defects move many interatomic spacings, chemical

impurities and dislocations may affect the process of annealing, by interacting with point defects. The crystal surface and dislocations may act as "sinks," on the approaching of which point defects may recombine. However, in many cases, including germanium and silicon single crystals, the dimensions are sufficiently great compared with the distances traveled by defects during annealing, and the density of dislocations is so small that the interaction of defects with one another and with impurity atoms predominates over other interactions.

Germanium. The results of an investigation of the annealing of radiation defects, generated by bombardment with 3 MeV electrons, were reported by Brown, Fletcher, and Wright [72]; they investigated only those effects which were stable at room temperature. As a measure of the defect concentration they used changes in the electron density in the conduction band, introducing, for convenience, the dimensionless quantity

$$f(t) = \frac{\sigma_0 - \sigma(t)}{\sigma_0 - \sigma_{AB}}, \qquad (5.47)$$

where σ_0 was the initial electrical conductivity at room temperature, σ_{AB} the electrical conductivity after bombardment and the formation of defects, and $\sigma(t)$ the electrical conductivity after annealing had proceeded for a time t. The use of the quantity $f(t)$, defined as the "fraction of the residual change in the electrical conductivity," has made it possible to represent annealing as the true recombination of point defects, which also destroys the energy levels of these defects in the forbidden bands. If these defects combine to form complexes (for example, divacancies) or if they combine with impurities, or if one, the more mobile, type of defect disappears on its approach to dislocations, only the less mobile defects will remain in a crystal. Then the simple interpretation of the electrical-conductivity changes on annealing as being due to the reestablishment of the initial structure is erroneous. Typical annealing curves, representing changes in the electrical conductivity of n-type germanium, are shown in Fig. 88 [67]. They indicate that the annealing is not a process of the first kind, which one would expect for closely spaced Frenkel pairs. Data of the type shown in Fig. 88 do not allow us to determine which of the point defects is more mobile. On the other hand, it is possible to determine the activation energy of jumps from one

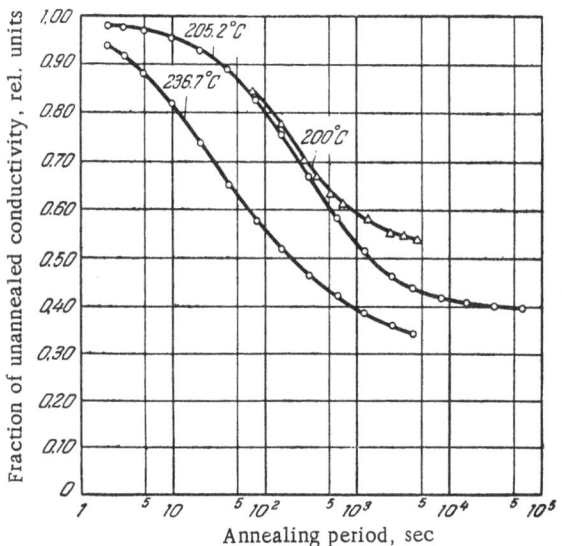

Fig. 88. Reestablishment of the electrical conductivity
due to the annealing of n-type germanium previously
irradiated with fast electrons [67, 72].

lattice site to another. In the case of n-type single crystals of
germanium, this energy is close to 1.4 eV.

A detailed review of the experiments on the annealing of radiation defects in germanium, carried out up to 1959, was given
by Brown, Augustyniak, and Waite [67]. These reviewers are of
the opinion that the reestablishment of the initial properties of irradiated germanium cannot be described by the simple model just
considered, although the mechanism of "discrete" diffusion of
defects, represented by a single activation energy, describes correctly the motion of these defects in a crystal. The annealing
processes are sometimes affected by the charge state of the moving
defects; it has been found that the annealing depends on the Fermi-
level position and that the illumination of a sample, causing photoionization of defects, affects the rate of annealing. In contrast
to the simple model represented by Fig. 87, an additional potential barrier may appear between a vacancy and an interstitial atom
due to the interaction of charges of each of these point defects, and
this affects the probability of their recombination.

Silicon. Detailed investigations of the annealing of radiation defects in silicon, using measurements of the conductivity and Hall effect, have not yet been carried out. As pointed out earlier, the reestablishment of the initial state did not occur at temperatures up to 150°C in the case of crystals irradiated at room temperature. However, it has been noticed that if crystals were irradiated at low temperatures, the subsequent heating to 70-300°K reestablished the initial structure [28].

To investigate high-temperature annealing of radiation defects in silicon, Bemski and Augustyniak [74] used measurements of the lifetime of nonequilibrium carriers. The results of measurements obtained at various annealing temperatures for a series of identical samples subjected to electron bombardment all fitted a single curve representing the dependence of a quantity proportional to the relative number of unannealed recombination centers on the logarithm of the "equivalent annealing period." An analysis of this curve showed that the process of annealing had a single activation energy of 1.3 eV. Bemski and Augustyniak concluded that the annealing rate was exponential, becoming bimolecular in the last stage. They related the process of the first kind (the exponential process) to the interaction between defects and electrically inactive centers present before irradiation.

The reestablishment of the carrier lifetime as a result of the annealing of electron-bombarded silicon was practically identical in n- and p-type crystals. According to the available data, the presence of copper, introduced deliberately into the silicon before irradiation, increased the relative importance of the bimolecular process, which might be accounted for by changes in the radius of a sphere in which one of the defects might be captured by a defect of the opposite type. A very high density of dislocations (more than 10^5 cm^{-2}) increased the rate of reestablishment of the initial state, other conditions being equal, by a factor of 100. A comparison of the data on the reestablishment of the lifetime by annealing with the data on electron-spin resonance of a center responsible for a recombination level at $E_c - 0.16$ eV [35, 36] showed that the initial value of the lifetime of nonequilibrium carriers was reestablished after the disappearance of these particular centers.

§ 32. Effects of Nuclear Reactions in Semiconductors

The irradiation of crystals with heavy charged particles and neutrons may give rise to nuclear reactions which produce chemical elements not initially present in the crystal. This important point was first mentioned by Lark-Horovitz [4], who analyzed the effects of nuclear reactions involving germanium isotopes and the influence of the resultant chemical impurities on the electrical properties of the irradiated crystals. At the time, Lark-Horovitz was taking part in experiments being carried out at the Oak Ridge Laboratory on the irradiation of germanium with slow neutrons. The results of the experiments were in good agreement with the calculations based on the data for neutron cross sections of germanium isotopes. Table 12, taken from the work reported here, lists the most important nuclear reactions involving germanium isotopes. It should be noted that data are available not only on the total activation cross section of germanium for thermal neutrons (this cross section is 2.3×10^{-24} cm^2) but also on the cross sections of individual isotopes (Table 13). The latter table shows that nuclear transformations in germanium irradiated with slow neutrons produce gallium (an acceptor) and arsenic (a donor) and that it is possible to calculate the difference between their concentrations, i.e., the difference between acceptor and donor concentrations.

For each of the nuclear transformations, the concentration of the resultant new atoms N_i is calculated from a simple formula

$$N_i = (n_n vt)\, \bar{\sigma}_i N_{Ge} P_i,$$

where $n_n vt = \Phi$ is the integral neutron flux, $\bar{\sigma}_i$ is the capture cross section, N_{Ge} is the number of germanium atoms in 1 cm^3, and P_i is the relative abundance of a given isotope (%).

In the experiments of Lark-Horovitz and others, germanium single crystals irradiated with slow neutrons were subjected to 24 hr annealing at 450°C (longer annealing produced no further changes in electrical conductivity). The purpose of the annealing was to remove the structural radiation defects produced by the action of gamma rays in the reactor. The precision of the results of Lark-Horovitz made it possible to confirm the hypothesis that each atom of group V contributes only one electron to the conduc-

TABLE 12. Nuclear Reactions in Germanium

Isotope abundance, %	Isotope
21.2	Ge^{70}
27.3	Ge^{72}
7.9	Ge^{73}
37.1	Ge^{74}
6.5	Ge^{76}

I. Principal reactions due to deuterons and slow neutrons

(d,p) (n,γ)

- $Ge^{70} \rightarrow Ge^{71} \xrightarrow{11.4\ days} Ga^{71}$
- $Ge^{72} \rightarrow Ge^{73}$
- $Ge^{73} \rightarrow Ge^{74}$
- $Ge^{74} \rightarrow Ge^{75} \xrightarrow{82\ min} As^{75}$
- $Ge^{76} \rightarrow Ge^{77} \xrightarrow{12\ hr,\ 59\ sec} As^{77} \xrightarrow{40\ hr} Se^{77}$

(d,n)

- $As^{71} \xrightarrow{50\ hr} Ge^{71} \xrightarrow{11\ days} Ga^{71}$
- $As^{73} \xrightarrow{76\ days} Ge^{73}$
- $As^{74} \xrightarrow{17.5\ days} {Ge^{74} \atop Se^{74}}$
- As^{75}
- $As^{77} \xrightarrow{40\ hr} Se^{77}$

II. Principal reactions due to fast neutrons

(n,p)

- $Ge^{70} \xrightarrow{20\ min} Ga^{70}$
- $Ge^{72} \xrightarrow{14.3\ hr} Ga^{72}$
- $Ge^{73} \xrightarrow{5\ hr} Ga^{73}$
- $Ge^{74} \xrightarrow{?} Ga^{74}$
- $Ge^{76} \xrightarrow{?} Ga^{76}$

(n,2n)

- $Ge^{70} \rightarrow Ge^{69} \xrightarrow{40\ hr} Ga^{69}$
- $Ge^{72} \rightarrow Ge^{71} \xrightarrow{11.4\ days} Ga^{71}$
- $Ge^{73} \rightarrow Ge^{72}$
- $Ge^{74} \rightarrow Ge^{73}$
- $Ge^{76} \rightarrow Ge^{75} \xrightarrow{82\ min} As^{75}$

(n,α)

- Zn^{67}
- $Zn^{69} \xrightarrow{14\ hr,\ 52\ min} Ga^{69}$
- Zn^{70}
- $Zn^{71} \xrightarrow{2.2\ min} Ga^{71}$
- $Zn^{73} \xrightarrow{<\ 2\ min} Ga^{73} \xrightarrow{5\ hr} Ge^{73}$

III. Principal reactions due to α particles

(α,n)

- $Ge^{70} \xrightarrow{7.1\ hr} Se^{73}$
- $Ge^{72} \xrightarrow{127\ days} Se^{75}$
- $Ge^{73} \rightarrow Se^{76}$
- $Ge^{74} \rightarrow Se^{77}$
- $Ge^{76} \xrightarrow{?} Br^{79}$

(α,2n)

- $Se^{73} \xrightarrow{76\ days} As^{73} \xrightarrow{9.7\ days} Se^{72} \xrightarrow{26\ hr} Ge^{72}$
- Se^{74}
- $Se^{75} \xrightarrow{127\ days} As^{75}$
- Se^{76}
- Se^{78}

(α,p)

- $As^{73} \xrightarrow{76\ days} Ge^{73}$
- $As^{75} \rightarrow Ge^{76}$
- $As^{76} \xrightarrow{26.8\ hr} {Ge^{76} \atop Se^{76}}$
- $As^{77} \xrightarrow{40\ hr} Se^{77}$
- $As^{79} \xrightarrow{?} Se^{79} \xrightarrow{?} Br^{79}$

TABLE 13. Cross Sections for Capture of Slow Neutrons by Germanium Isotopes

Isotope	Abundance, %	Capture cross section, 10^{-24} barns		Stable element
		for nuclide	for atom	
Ge^{70}	21.2	3.25	0.69	Ga
Ge^{72}	27.3	0.94	0.26	Ge
Ge^{73}	7.9	13.69	1.08	Ge
Ge^{74}	37.1	0.60	0.22	As
Ge^{76}	6.5	0.35	0.02	Se

tion band, in germanium, and each atom of group III captures an electron, giving rise to a hole. At the time of these experiments (1951) this revelation was of paramount importance to the physics of semiconductors. A similar technique was later used by Kekelidze [73] to check the method of determining separately the donor and acceptor concentrations in germanium from measurements of the Hall effect at low temperatures.

Although quantitative data on the generation of chemical impurities by neutron irradiation are at present available only for germanium, one can expect that this method of introducing local centers will find application in subsequent investigations and the practical utilization of other semiconductors, in particular, semiconducting compounds.

H. Schweinler [75], as well as M. V. Chukichev and V. S. Vavilov [76], have pointed out that the recoil of nuclei produced by neutron capture might be accompanied by the generation of a large number of defects because such nuclei emit gamma quanta. In particular, in the case of the irradiation of silicon in reactors, the number of defects products by the recoil is comparable [76] with the number of defects produced by the scattering of fast neutrons. This effect is particularly strong in semiconductors containing atoms of low atomic weight A, since the nuclear recoil energy is given in the form

$$E_A = \frac{537}{A} (h\nu)^2, \tag{5.48}$$

where $h\nu$ is the energy of the emitted gamma quantum. Calculations of this effect require complete and reliable data on the transitions in excited nuclei. Such data can be found, for example, in the atlas of L. V. Groshev et al. [77].

LITERATURE CITED

Preface

1. V. K. Subashiev and M. S. Sominskii, "Semiconducting photo-cells," in collection: Semiconductors in Science and Technology, Vol. II, A. F. Ioffe (ed.); Izd. Akad. Nauk SSSR, 1958, pp. 115-217.
2. V. S. Vavilov, "Semiconducting converters of radiation energy," Usp. Fiz. Nauk 56:111 (1955).
3. "Semiconducting converters of radiation energy" [Russian translation], in collection ed. by Yu. P. Maslakovets and V. K. Subashiev, IL, 1959.
4. R. Newman and W. W. Tyler, "Photoconductivity of germanium," Usp. Fiz. Nauk 72:587 (1960).
5. R. A. Smith, F. E. Jones, and R. P. Chasmar, Detection and Measurement of Infrared Radiation [Russian translation], IL, 1959, Chapts. IV–VII.
6. N. G. Basov, O. N. Krokhin, and Yu. M. Popov, "Generation, amplification, and display of infrared and visible radiation using quantum systems," Usp. Fiz. Nauk 72:161 (1960).
7. G. Troup, Masers, Microwave Amplification, and Oscillation by Stimulated Emission [Russian translation], IL, 1961.
8. A. Shavlov, "Optical masers," Usp. Fiz. Nauk 75:569 (1961).
9. Proceedings of a Symposium on Semiconducting Detectors of Nuclear Radiation (held in USA), IRE Transactions on Nuclear Science, N S-8, No. 1, 1961.

Chapter I

1. H. Y. Fan, "Infrared absorption by semiconductors," Usp. Fiz. Nauk 64:316 (1958).
2. H. Rank, Phys. Rev. 90:202 (1953).
3. H. Briggs, Phys. Rev. 78:287 (1958).
4. N. B. Hannay, Semiconductors [Russian translation], IL, 1962, Chapt. X.

5. N. F. Mott and R. W. Gurney, Electronic Processes in Ionic Crystals [Russian translation], IL, 1950, p. 185.

6. W. Dash and R. Newman, Phys. Rev. 99:1151 (1955); see also [1] above.

7. F. Herman, Phys. Rev. 95:847 (1954).

8. J. Bardeen, F. Blatt, and L. Hall, Proceedings of a Conference on Photoconductivity, John Wiley and Sons, Inc., New York, 1954.

9. O'Brien, J. Opt. Soc. Am. 26:122 (1936).

10. W. Scanlon, J. Phys. Chem. Solids 8:423 (1959).

11. D. Bell et al., Proc. Roy. Soc. (London) A217:71 (1953).

12. G. Parkinson et al., Proc. Phys. Soc. (London) B67:644 (1954).

13. M. Tanenbaum and H. Briggs, Phys. Rev. 91:1561 (1953).

14. G. Kimball, J. Chem. Phys. 3:560 (1935); see also F. Seitz, Modern Theory of Solids [Russian translation], IL, 1949, p. 480.

15. H. Y. Fan, Phys. Rev. 78:808 (1950).

16. H. Y. Fan, M. L. Shepherd, and W. Spitzer, Proceedings of a Conference on Photoconductivity, John Wiley and Sons, Inc., New York, 1954.

17. T. S. Moss, Optical Properties of Semiconductors [Russian translation], IL, 1961.

18. J. R. Haynes, M. Lax, and W. Flood, J. Phys. Chem. Solids 8:394 (1959).

19. W. Shockley and J. Bardeen, Phys. Rev. 77:407 (1950).

20. G. G. Macfarlane, T. P. McLean, J. E. Quarrington, and V. Roberts, J. Phys. Chem. Solids 8:388 (1959).

21. W. Spitzer et al., Phys. Rev. 98:228 (1955); D. Warshauer et al., Phys. Rev. 98:1193 (1955).

22. L. V. Keldysh, Zh. Eksperim. i Tech. Fiz. 34:1138 (1958).

23. W. Franz, Z. Naturforsch. 13a:484 (1958).

24. L. V. Keldysh, V. S. Vavilov, and K. I. Britsyn, Proceedings of the 2nd International Conference on Semiconductors, Prague, 1961, p. 824.

25. F. F. Vol'kenshtein, Tr. Fiz. Inst. Akad. Nauk SSSR 1:123 (1937).

26. V. S. Vavilov and K. I. Britsyn, Fiz. Tverd. Tela 2:1937 (1960).

27. H. Y. Fan and A. K. Ramdas, J. Appl. Phys. 30:1127 (1959).

28. V. S. Vavilov, A. F. Plotnikov, and G. V. Zakhvatkin, Fiz. Tverd. Tela 1:976 (1959).
29. K. I. Britsyn and V. S. Vavilov, Fiz. Tverd. Tela 3:2497 (1961).
30. S. Zwerdling et al., Phys. Rev. 108:1402 (1957).
31. F. Seitz, Modern Theory of Solids, McGraw-Hill Book Co., Inc., New York, 1940, p. 583.
32. L. Apker and E. Taft, Phys. Rev. 79:964 (1950); 81:698 (1950); 82:814 (1951).
33. E. F. Gross, Usp. Fiz. Nauk 63:576 (1957); J. Phys. Chem. Solids 8:172 (1959).
34. H. Y. Fan and P. Fischer, J. Phys. Chem. Solids 8:270 (1959).
35. C. Kittel and A. Mitchell, Phys. Rev. 96:1488 (1954); [Russian translation] in the collection: Problems of Semiconductor Physics, IL, 1957, p. 505.
36. W. Kohn and J. Luttinger, Phys. Rev. 98:915 (1955).
37. W. Kohn, Solid State Physics, Vol. 5, F. Seitz (ed.), Academic Press, New York, 1958, p. 257.
38. V. S. Vavilov, E. N. Lotkova, and A. F. Plotnikov, Photoconductivity, Pergamon Press, London, 1962, p. 31.
39. R. Newman and W. W. Tyler, Usp. Fiz. Nauk 72:587 (1960).
40. V. S. Vavilov, Fiz. Tverd. Tela 2:364 (1960).
41. R. Kronig, Proc. Roy. Soc. (London) A133:255 (1931).
42. A. H. Kane, Phys. Rev. 97:1647 (1955).
43. P. Wolfe, Proc. Phys. Soc. (London) A67:74 (1954).
44. P. Dexter and B. Lax, Phys. Rev. 96:223 (1954).
45. V. I. Zvyagin and V. S. Vavilov, Pribory i Tekhn. Eksperim. 1:86 (1956).
46. V. A. Yakovlev, Fiz. Tverd. Tela 2:1624, 2639 (1960).
47. H. B. Briggs and R. C. Fletcher, Phys. Rev. 91:1342 (1953).
48. N. J. Harrick, Phys. Rev. 101:491 (1956).
49. K. Lehovec, Proc. Inst. Radio Engrs. 40:1407 (1952).
50. Yu. I. Ukhanov, Doklady Akad. Nauk SSSR 111:1238 (1956).
51. F. Stern, Phys. Rev. 108:158 (1957).
52. A. K. Ramdas and H. Y. Fan, Bull. Am. Phys. Soc. 3:121 (1958).
53. W. Kaiser and P. Keck, J. Appl. Phys. 28:882 (1957).
54. H. Hrostowski and R. Kaiser, Phys. Rev. 107:966 (1957).
55. W. Kaiser, Phys. Rev. 105:1751 (1957).

56. B. Szigeti, Proc. Roy. Soc. (London) A204:52 (1950).
57. C. Kittel, Introduction to Solid State Physics [Russian transla-
tion], Gostekhizdat, 1957, Chapt. IV.
58. R. Collins and H. Y. Fan, Phys. Rev. 93:674 (1954).
59. M. Lax and E. Burstein, Phys. Rev. 97:39 (1955).

Chapter II

1. S. M. Ryvkin, "Recombination in semiconductors," in collection: Semiconductors in Science and Technology, Vol. II,
Izd. Akad. Nauk SSSR, 1958, p. 463.
2. R. H. Bube, Photoconductivity of Solids [Russian translation],
IL, 1962, Chapts. III, X, XI.
3. B. M. Vul, Fiz. Tverd. Tela 3:2264 (1961).
4. B. Gudden, Lichtelektrische Erscheinungen, Berlin, 1928.
5. P. S. Tartakovskii, Internal Photoeffect in Dielectrics, Gostekhizdat, 1940.
6. V. S. Vavilov, "Radiation ionization processes in germanium
and silicon crystals," Usp. Fiz. Nauk 75:263 (1961).
7. W. Shockley, "The problem of p−n junctions in silicon," Usp.
Fiz. Nauk, May, 1962.
8. S. M. Ryvkin, Doklady Akad. Nauk SSSR 72(3):481 (1950).
9. F. S. Goucher, Phys. Rev. 78:816 (1950).
10. N. F. Mott and R. W. Gurney, Electronic Processes in Ionic Crystals [Russian translation], IL, 1950, Chapts. III, IV.
11. J. Tauc, J. Phys. Chem. Solids 8:219 (1958).
12. V. K. Subashiev and M. S. Sominskii, "Semiconducting photo-
cells," in collection: Semiconductors in Science and Technology, Vol. II, Izd. Akad. Nauk SSSR, 1958, p. 115.
13. W. Pfann and W. Van Roosbroeck, J. Appl. Phys. 25:1422
(1954).
14. V. S. Vavilov, V. M. Malovetskaya, and G. N. Galkin,
Atomnaya Energ. 4:571 (1958).
15. S. I. Vavilov, "Luminescence and its duration," in: Collected
Works, Vol. II, Izd. Akad. Nauk SSSR, 1952, p. 293.
16. F. A. Butaeva and V. A. Fabrikant, Izv. Akad. Nauk SSSR,
Ser. Fiz. 21:541 (1957).
17. S. Koc, Cesk. Casopis Fys. 6:668 (1956).
18. V. S. Vavilov, J. Phys. Chem. Solids 8:223 (1959).
19. K. I. Britsyn and V. S. Vavilov, Opt. i Spektroskopiya 8:861
(1960).

20. V. S. Vavilov and K. I. Britsyn, Zh. Eksperim. i Teor. Fiz. 34:1354 (1958).
21. H. Y. Fan, Usp. Fiz. Nauk 64:315 (1958).
22. V. M. Patskevich, V. S. Vavilov, and L. S. Smirnov, Zh. Eksperim. i Teor. Fiz. 33:804 (1957).
23. K. I. Britsyn, Dissertation, MGU, 1961.
24. K. G. McKay, Phys. Rev. 108:20 (1957).
25. P. Wolfe, Phys. Rev. 95:1415 (1954).
26. V. S. Vavilov, Doctoral dissertation, Effect of Radiations on Germanium and Silicon, Fiz. Inst. im. P. N. Lebedeva Akad. Nauk SSSR, Moscow, 1960; see also [6] above.
27. E. Antoncik, Cesk. Casopis Fys. 7:651 (1957).
28. J. Drahokoupil, M. Malkovska, and J. Tauc, Cesk. Casopis Fys. 7:521 (1957).
29. M. V. Chukichev and V. S. Vavilov, Fiz. Tverd. Tela 3:935 (1961).
30. F. Herman, J. Phys. Chem. Solids 2:72 (1957).
31. N. Sclar and E. Burstein, Phys. Rev. 98:1757 (1955).
32. H. Gummel and M. Lax, Phys. Rev. 97:1469 (1955).
33. R. H. Hall, Phys. Rev. 87:387 (1952).
34. W. Shockley and W. Read, Phys. Rev. 87:835 (1953).
35. J. Burton et al., J. Phys. Chem. 57:853 (1953).
36. S. G. Kalashnikov, Proceedings of Second International Conference on Semiconductors, 1960, Prague, 1961, p. 241.
37. W. Shockley and C.-T. Sah, Phys. Rev. 109:1103 (1958).
38. S. G. Kalashnikov, Fiz. Tverd. Tela 2:2743 (1960).
39. S. G. Kalashnikov and K. P. Tissen, Fiz. Tverd. Tela 1:1754 (1959).
40. T. Okada, J. Phys. Soc. Japan 10:1110 (1955).
41. S. Kulin et al., J. Appl. Phys. 27:1287 (1956).
42. N. B. Hannay, Semicondcutors [Russian translation], IL, 1962.
43. T. S. Moss, Optical Properties of Semiconductors [Russian translation], IL, 1961, Chapt. XVI.
44. G. Wertheim, Phys. Rev. 104:662 (1956).
45. J. R. Haynes and J. Hornbeck, in collection: Problems of Semiconductor Physics [Russian translation], IL, 1957, pp. 167, 187.
46. R. Shulman, Phys. Rev. 102:1451 (1956).
47. R. Shulman and B. Wyluda, Phys. Rev. 102:1455 (1956).

48. N. B. Hannay, Semiconductors [Russian translation], IL, 1962, Chapt. XI.
49. H. Hrostowski and R. Kaiser, Phys. Rev. 107:966 (1957).
50. N. B. Hannay, Semiconductors [Russian translation], IL, 1962, Chapt. XVI.
51. C. G. B. Garrett and W. H. Brattain, Phys. Rev. 99:376 (1955).
52. A. Rose, in collection: Problems of Semiconductor Physics [Russian translation], IL, 1957, p. 130.
53. R. H. Bube, Photoconductivity of Solids [Russian translation], IL, 1962, Chapt. XI.
54. B. Borshchevskii, J. Phys. Chem. USSR 21:1007 (1947).
55. E. G. Miselyuk and E. B. Mertens, Izv. Akad. Nauk SSSR, Ser. Fiz. 16:115 (1952).
56. M. Borisov and S. Kaniev, Z. Physik. Chem. 205:56 (1955).
57. F. Stockmann, Z. Physik 143:348 (1955).
58. V. L. Bonch-Bruevich, Zh. Eksperim. i Teor. Fiz. 28:67 (1955).
59. R. Newman and W. W. Tyler, "Photoconductivity of germanium," Usp. Fiz. Nauk 72:587 (1960).
60. D. Wright, Brit. J. Appl. Phys. 9:205 (1958).
61. S. M. Ryvkin, Fiz. Tverd. Tela 2:2411 (1960).
62. V. P. Dobrego and S. M. Ryvkin, Fiz. Tverd. Tela 4:553 (1962).
63. L. S. Milevskii, Fiz. Tverd. Tela 4:429, 825 (1962).

Chapter III

1. E. Segrè (ed.), Experimental Nuclear Physics, Vol. I, Part II [Russian translation], IL, 1956, pp. 143-291.
2. N. Bohr, The Penetration of Atomic Particles through Matter [Russian translation], IL, 1950.
3. H. Bethe, Ann. Physik 5:325 (1930).
4. F. Seitz, Discussions Faraday Soc. 5:271 (1949).
5. A. R. Shul'man (ed.), Characteristic Electron Energy Losses [Russian translation], IL, 1959.
6. L. Spencer, Phys. Rev. 98:1597 (1955).
7. B. Ya. Yurkov, Zh. Tekhn. Fiz. 28:1159 (1958).
8. B. M. Vul, V. S. Vavilov, L. S. Smirnov, and G. N. Galkin, Atomnaya Energ. 2:533 (1957).

9. A. G. Chynoweth, "Crystal counters of conducting type, " in collection: Effects of Radiation on Semicondcutors and Insulators [Russian translation], IL, 1954, p. 160.

10. R. Hofstadter, Nucleonics, No. 4:2 (1949); No. 5:29 (1949); see also Usp. Fiz. Nauk 39:462 (1949).

11. W. Shockley, Electrons and Holes in Semiconductors [Russian translation], IL, 1953.

12. K. G. McKay, Phys. Rev. 84:829 (1951).

13. K. G. McKay and K. B. McAffe, Phys. Rev. 91:1079 (1953).

14. V. S. Vavilov, L. S. Smirnov, and V. M. Patskevich, Doklay Akad. Nauk SSSR 112:1020 (1957).

15. V. M. Patskevich, V. S. Vavilov, and L. S. Smirnov, Zh. Eksperim i Teor. Fiz. 33:804 (1957).

16. V. S. Vavilov, J. Phys. Chem. Solids 8:223 (1958).

17. V. S. Vavilov, V. M. Malovetskaya, and G. N. Galkin, Atomnaya Energ. 4:571 (1958).

18. P. Kennedy, Proc. Roy. Soc. (London) 253:37 (1959).

19. K. G. McKay, Phys. Rev. 77:817 (1950).

20. W. Shockley, Usp. Fiz. Nauk 77:161 (1962).

21. J. Tauc, J. Phys. Chem. Solids 8:219 (1958).

22. M. V. Chukichev and V. S. Vavilov, Fiz. Tverd. Tela 3:935 (1961).

23. J. Drahokoupil, M. Malkovska, and J. Tauc, Czech. J. Phys. 7:521 (1957).

Chapter IV

1. O. V. Losev, Telegrafiya i Telefoniya, No. 18:61 (1923); No. 26:403 (1924); No. 44:485 (1927); No. 53:153 (1929); Phil. Mag. 7:1024 (1928).

2. F. S. Goucher, Phys. Rev. 78:816 (1950).

3. W. Van Roosbroeck and W. Shockley, Phys. Rev. 94:1558 (1954); Russian translation in collection: Problems of Semiconductor Physics, IL, 1957, p. 122.

4. H. B. Briggs, Phys. Rev. 77:287 (1950).

5. E. Burstein and P. H. Egli, "Physics of semiconductors" (a review), in collection: Physics of Semiconductors [Russian translation], 1957.

6. E. I. Adirovich, Some Problems in the Theory of Luminescence of Crystals, Gostekhizdat, 1956, Chapt. VI.

7. V. L. Bonch-Bruevich (ed.), Collection: Problems of Semi-conductor Physics [Russian translation], IL, 1957, Chapt. IV.

8. J. R. Haynes, Phys. Rev. 98:1866 (1953).

9. J. R. Haynes and H. B. Briggs, Phys. Rev. 86:647 (1952).

10. P. Aigrain and C. Benôit à la Guillaume, J. Phys. Radium 17:709 (1956).

11. R. Newman, Phys. Rev. 91:1313 (1953).

12. R. Smith, Advan. Phys. 2:321 (1953); Russian translation in collection: Effect of Radiation on Semiconductors and Insulators, IL, 1954, p. 226.

13. H. Y. Fan, Usp. Fiz. Nauk 64:315 (1958).

14. J. R. Haynes, M. Lax, and W. Flood, J. Phys. Chem. Solids 8:392 (1959).

15. G. G. Macfarlane et al., Phys. Rev. 111:1245 (1958).

16. B. B. Brockhouse, J. Phys. Chem. Solids 8:400 (1959).

17. R. Newman, Phys. Rev. 105:1715 (1957).

18. A. Tweet, Phys. Rev. 99:1245 (1955).

19. V. S. Vavilov, A. A. Gippius, M. M. Gorshkov, and B. D. Kopylovskii, Zh. Eksperim. i Teor. Fiz. 37:23 (1959).

20. C. Benôit à la Guillaume, Compt. Rend. 243:704 (1956).

21. J. R. Haynes and W. Westphal, Phys. Rev. 101:1676 (1956).

22. F. Morin et al., Phys. Rev. 96:833 (1954).

23. E. Burstein et al., J. Phys. Chem. 57:849 (1953).

24. T. S. Moss and T. Hawkins, Phys. Rev. 101:1609 (1956).

25. T. S. Moss, Optical Properties of Semiconductors [Russian translation], IL, 1961, Chapts. VI, XVI.

26. N. G. Basov, B. D. Osipov, and A. N. Khvoshchev, Zh. Eksperim. i Teor. Fiz. 40:1880 (1961).

27. R. Braunstein, Phys. Rev. 99:1892 (1955).

28. E. Blount et al., Phys. Rev. 96:576 (1954).

29. H. Welker, Physica 20:893 (1954).

30. L. N. Galkin and N. V. Korolev, Doklady Akad. Nauk SSSR 92:529 (1953).

31. W. Scanlon, Phys. Rev. 109:47 (1958).

32. N. G. Basov, Yu. M. Popov, and O. N. Krokhin, Usp. Fiz. Nauk 72:161 (1960).

33. G. Troup, Masers, Microwave Amplification, and Oscillation by Stimulated Emission [Russian translation], IL, 1961.

34. N. G. Basov, B. M. Vul, and Yu. M. Popov, Zh. Eksperim. i Teor. Fiz. 37:587 (1959).

35. O. N. Krokhin, Dissertation, Fiz. Inst. Akad. Nauk SSSR, 1962.
36. L. V. Keldysh, Zh. Eksperim. i Teor. Fiz. 37:713 (1959).
37. B. Lax, Quantum Electronics, New York, 1960.
38. N. G. Basov, O. N. Krokhin, and Yu. M. Popov, Vest. Akad. Nauk SSSR, No. 3:61 (1961).
39. O. N. Krokhin, Fiz. Tverd. Tela 4:829 (1962).
40. L. Davies, Phys. Rev. Letters 4:11 (1960).
41. D. N. Nasledov, A. A. Rogachev, S. M. Ryvkin, and B. V. Tsarenkov, Fiz. Tverd. Tela 4:1062 (1962).
42. R. N. Hall, G. E. Fenner, J. D. Kingsley, T. J. Soltys, and R. O. Carlson, Phys. Rev. Letters 9:336 (1962).
43. V. S. Bagaev et al., Doklady Akad. Nauk SSSR 150:67 (1963).

Chapter V

1. F. Seitz, in collection: Effect of Radiation on Semiconductors and Insulators [Russian translation], S. M. Ryvkin (ed.), IL, 1954, p. 9.
2. F. Seitz and J. Koehler, Solid State Physics, Vol. II, Academic Press, New York, 1956, pp. 307-442.
3. W. McKinley, Phys. Rev. 74:1759 (1948).
4. K. Lark-Horovitz, in collection: Semiconducting Materials [Russian translation], V. M. Tuchkevich (ed.), IL, 1954, pp. 62-95.
5. J. Kahn, J. Appl. Phys. 30:1310 (1959).
6. L. Katz and A. Penfold, Rev. Mod. Phys. 24:28 (1952).
7. G. Kinchin and R. Pease, Rept. Progr. Phys. Vol. 18 (1955); see also Usp. Fiz. Nauk 60:590 (1956).
8. R. Van de Graaff et al., Phys. Rev. 69:452 (1946).
9. W. Kohn, Phys. Rev. 94:1409 (1954).
10. H. Smith, Phil. Trans. Roy. Soc. (London) A241:105 (1948).
11. R. Fletcher and W. Brown, Phys. Rev. 92:585 (1953).
12. W. Brown and W. Augustyniak, J. Appl. Phys. 30:1300 (1959).
13. L. S. Smirnov and P. Ya. Glazunov, Fiz. Tverd. Tela 1:1376 (1959).
14. V. V. Galavanov, Fiz. Tverd. Tela 1:432 (1959).
15. O. Oen and D. Holmes, J. Appl. Phys. 30:1289 (1959).
16. G. Dienes and G. Vineyard, Radiation Damage in Solids [Russian translation], IL, 1960, Chapt. III.

17. W. Snyder and J. Neufeld, Phys. Rev. 97:1636 (1955).
18. J. Sampson, Phys. Rev. 99:1657 (1955).
19. G. Dienes and G. Vineyard, Radiation Damage in Solids, Interscience Publ., Inc., New York, 1957.
20. A. F. Ioffe, Physics of Semiconductors, Izd. Akad. Nauk SSSR, 1957, p. 332.
21. J. Brinkman, J. Appl. Phys. 25:961 (1954).
22. H. Brooks, J. Appl. Phys. 30:1118 (1959).
23. H. James and K. Lark-Horovitz, Z. Physik. Chem. (Leipzig) 198:107 (1951).
24. H. Y. Fan and K. Lark-Horovitz, in collection: Physics of Semiconductors [Russian translation], V. S. Vavilov (ed.), IL, 1957, p. 161.
25. S. M. Ryvkin, L. G. Paritskii, R. Yu. Khansevarov, and I. D. Yaroshetskii, Fiz. Tverd. Tela 3:252 (1961).
26. J. Loferski and P. Rappaport, Phys. Rev. 111:432 (1958).
27. V. S. Vavilov, V. M. Patskevich, B. Ya. Yurkov, and P. Ya. Glazunov, Fiz. Tverd. Tela 2:1431 (1960).
28. G. Wertheim, Phys. Rev. 105:1730 (1957); 110:1272 (1958).
29. D. Hill, Phys. Rev. 114:1414 (1959).
30. B. Abeles and S. Meiboom, Phys. Rev. 95:31 (1957).
31. V. Johnson and K. Lark-Horovitz, Phys. Rev. 82:977 (1951).
32. F. Shipley and V. Johnson, Phys. Rev. 90:523 (1953).
33. H. Y. Fan, Solid State Physics, Vol. I, Academic Press, New York, 1955, p. 288.
34. G. M. Galkin, N. S. Rytova, and V. S. Vavilov, Fiz. Tverd. Tela 2:2025 (1960).
35. G. Bemski, J. Appl. Phys. 30:1195 (1959).
36. G. Watkins et al., J. Appl. Phys. 30:1198 (1959).
37. H. Y. Fan and A. K. Ramdas, J. Appl. Phys. 30:1127 (1959).
38. V. S. Vavilov, E. N. Lotkova, and A. F. Plotnikov, Proceedings of the Second International Conference on Photoconductivity, Pergamon Press, London, 1962, p. 31.
39. M. Becker, H. Y. Fan, and K. Lark-Horovitz, Phys. Rev. 85:730 (1952).
40. V. S. Vavilov, A. F. Plotnikov, and G. V. Zakhvatkin, Fiz. Tverd. Tela 1:976 (1959).
41. B. Brockhouse, J. Phys. Chem. Solids 8:400 (1959).
42. V. S. Vavilov and A. F. Plotnikov, Fiz. Tverd. Tela 3:2455 (1961).

43. S. G. Kalashnikov and N. A. Penin, Zh. Tekhn. Fiz. 25:111 (1955).

44. G. Wertheim, Phys. Rev. 111:1500 (1958).

45. G. Wertheim, J. Appl. Phys. 30:1166 (1959).

46. V. S. Vavilov, L. S. Smirnov, G. N. Galkin, A. V. Spitsyn, and V. M. Patskevich, Zh. Tekhn. Fiz. 26:1865 (1956); 28:960 (1958).

47. J. Cleland, J. Crawford, and J. C. Pigg, Phys. Rev. 98:1742 (1955); 99:1170 (1955).

48. W. Spear, Phys. Rev. 112:362 (1958).

49. G. C. Messenger and J. P. Spratt, Proc. Inst. Radio Engrs. 46:1039 (1958).

50. V. S. Vavilov, L. S. Smirnov, A. V. Spitsyn, and M. V. Chukichev, Zh. Eksperim. i Teor. Fiz. 32:702 (1957).

51. A. V. Spitsyn and V. S. Vavilov, Zh. Eksperim. i Teor. Fiz. 34(2) (1958).

52. O. Curtis and J. Cleland, Bull. Am. Phys. Soc., Ser. II 4:47 (1959).

53. O. Curtis, J. Cleland, and J. Crawford, J. Appl. Phys. 29:1722 (1958).

54. S. G. Kalashnikov, Zh. Tekh. Fiz. 26:241 (1956).

55. R. Hasiguti et al., J. Phys. Soc. Japan 12:1351 (1957).

56. J. Cleland, J. Crawford, and D. Holmes, Phys. Rev.102:722 (1956).

57. V. S. Vavilov and L. S. Smirnov, Zh. Tekhn. Fiz. 27:427 (1957).

58. L. G. Sharendo and L. S. Smirnov, Fiz. Tverd. Tela 4:2137 (1962).

59. P. Baruch, J. Phys. Chem. Solids 8:153 (1959).

60. S. M. Ryvkin, R. Yu. Khansevarov, and I. D. Yaroshetskii, Fiz. Tverd. Tela 3:3211 (1961).

61. V. S. Vavilov, A. A. Gippius, M. M. Gorshkov, and B. D. Kopylovskii, Zh. Eksperim. i Teor. Fiz. 37:23 (1959).

62. R. Shulman, Phys. Rev. 102:1451 (1956).

63. J. Cleland and J. Crawford, Phys. Rev. 96:1177 (1954).

64. L. Aukerman, J. Appl. Phys. 30:1239 (1959).

65. J. Cleland and J. Crawford, Bull. Am. Phys. Soc. 3:142 (1958).

66. L. Aukerman and K. Lark-Horovitz, Bull. Am. Phys. Soc. 1:332 (1956).

67. W. Brown, W. Augustyniak, and T. Waite, J. Appl. Phys. 30:1258 (1959).
68. J. Corbett and R. Walker, Phys. Rev. 110:767 (1958).
69. J. W. MacKay and E. E. Klontz, J. Appl. Phys. 30:1269 (1959).
70. V. V. Antonov-Romanovskii, J. Phys. USSR 6:120 (1942); 7:153 (1943); Opt. i Spektroskopiya 3:592 (1957).
71. T. Waite, Phys. Rev. 107:471 (1957).
72. W. Brown, R. Fletcher, and D. Wright, Phys. Rev. 92:591 (1953).
73. N. P. Kekelidze, Proceedings of a Symposium of the International Atomic Agency on Chemical Effects of Nuclear Interactions, Prague, October 1960.
74. G. Bemski and W. Augustyniak, Phys. Rev. 108:645 (1957).
75. H. Schweinler, J. Appl. Phys. 30:1125 (1959).
76. M. V. Chukichev and V. S. Vavilov, Fiz. Tverd. Tela 3:1522 (1961).
77. L. V. Groshev et al., Atlas of Gamma-Ray Spectra due to Radiative Capture of Thermal Neutrons, 1958.
78. L. S. Smirnov, Fiz. Tverd. Tela 2:1669 (1960).
79. V. S. Vavilov, G. N. Galkin, V. M. Malovetskaya, and A. F. Plotnikov, Fiz. Tverd. Tela 4:1969 (1962).
80. J. Loferski and P. Rappaport, J. Appl. Phys. 30:1296 (1959).

INDEX